Capturing Value in Digital Health Eco-Systems
Validating Strategies for Stakeholders

Felix Lena Stephanie
School of Mechanical and Aerospace Engineering
Nanyang Technological University, Singapore

Ravi S. Sharma
Professor of Technological Innovation
Zayed University, Abu Dhabi Campus, United Arab Emirates

CRC Press is an imprint of the
Taylor & Francis Group, an **informa** business

A SCIENCE PUBLISHERS BOOK

First edition published 2022
by CRC Press
6000 Broken Sound Parkway NW, Suite 300, Boca Raton, FL 33487-2742

and by CRC Press
2 Park Square, Milton Park, Abingdon, Oxon, OX14 4RN

© 2022 Taylor & Francis Group, LLC

CRC Press is an imprint of Taylor & Francis Group, LLC

Reasonable efforts have been made to publish reliable data and information, but the author and publisher cannot assume responsibility for the validity of all materials or the consequences of their use. The authors and publishers have attempted to trace the copyright holders of all material reproduced in this publication and apologize to copyright holders if permission to publish in this form has not been obtained. If any copyright material has not been acknowledged please write and let us know so we may rectify in any future reprint.

Except as permitted under U.S. Copyright Law, no part of this book may be reprinted, reproduced, transmitted, or utilized in any form by any electronic, mechanical, or other means, now known or hereafter invented, including photocopying, microfilming, and recording, or in any information storage or retrieval system, without written permission from the publishers.

For permission to photocopy or use material electronically from this work, access www.copyright.com or contact the Copyright Clearance Center, Inc. (CCC), 222 Rosewood Drive, Danvers, MA 01923, 978-750-8400. For works that are not available on CCC please contact mpkbookspermissions@tandf.co.uk

Trademark notice: Product or corporate names may be trademarks or registered trademarks and are used only for identification and explanation without intent to infringe.

ISBN: 978-1-032-12339-4 (hbk)
ISBN: 978-1-032-12342-4 (pbk)
ISBN: 978-1-003-22415-0 (ebk)

Typeset in Times New Roman
by Innovative Processors

The first author makes a special dedication to her father

Dr. Felix Paul

a cancer patient, for his endless support and encouragement throughout her research journey.

Preface

It is laudable that one of the Sustainable Development Goals of the United Nations seeks "to ensure healthy lives and promote well-being for all at all ages". As a strategic roadmap for this goal, the World Health Organisation has undertaken various initiatives including establishing guidelines and recommendations on digital health intervention, applications, assessment, implementation and evaluation.

Regrettably, as the Covid-19 pandemic has revealed, much work remains to be done especially at the eco-system level, to achieve the lofty goal of universal health coverage. It is the conviction of the authors that a digital health eco-system with fairness and efficiency designed into it will be able to further the cause of equitable healthcare access to all. The authors dedicate this book to the well-being of future generations who will seek digital healthcare as a matter of universal right.

Overview of the Book

The United Nations' Sustainable Development Goals call for the establishment of Good Health and Well-being and targets a universal digital healthcare ecosystem by 2030. However, existing technology infrastructure is ineffectual to support the envisioned target and requires massive reconfiguration to achieve its intended outcome. This book suggests a way forward with fair and efficient digital health networks that provide resource efficiencies and inclusive access to those who are currently under-served. Specifically, a fair and efficient digital health network that provides a common platform to its key stakeholders to facilitate sharing of information with a view to promoting cooperation and maximizing benefits. A promising platform for this critical application is 'cloud technology' with its offer of computing as a utility and resource sharing. This is an area that has attracted much scholarly attention as it is well-suited to foster such a network bringing together diverse players who would otherwise remain fragmented and be unable to reap benefits that accrue from cooperation. The fundamental premise is that the notion of value in a digital-health ecosystem is brought about by digital information sharing and exchange.

Notwithstanding the potential of information and communication technology to transform the healthcare industry for the better, there are several barriers to its adoption, the most significant one being misaligned incentives for some stakeholders. This monograph suggests among other findings, that e-health in its true sense can become fair and efficient if a trusted third party assumes responsibility as the custodian of its citizens' health information so that 'collaboration for value' will replace 'competition for revenue' as the new axiom in delivering the public, good healthcare, through digital networks.

List of Tables

Table 2.1. Benefits of Blockchain for Digital Healthcare (Agbo et al., 2019) 30
Table 2.2. PHC Initiatives around the globe 48
Table 3.1. Revenue sources for E-Health infomediaries 60
Table 3.2. Vocabulary of research concepts 61
Table 3.3. Phase 1: List of target informants 68
Table 3.4. Phase 2: List of target informants 73
Table 4.1. Key players in E-Health: Concept matrix based on literature review and industry observations 83
Table 4.2. Key players in the E-Health ecosystem and their roles 86
Table 4.3. Analysis of the E-health ecosystem using ADVISOR 91
Table 4.4. Validated values created vs. values captured 96
Table 7.1. WHO statistics on healthcare indicators for 2014 141
Table A.1. Primary healthcare market players 162
Table A.2. Value created vs. value captured 163

List of Figures

Figure 1.1.	Research design	7
Figure 2.1.	Chapter framework of emerging information technologies for digital health.	13
Figure 2.2.	An overview of IoT enabled personalized healthcare (Wan et al., 2018, p. 2).	14
Figure 2.3.	A generic overview of the IoT paradigm in healthcare with emerging technologies (Khezr et al., 2019, p. 15)	15
Figure 2.4.	A conceptual representation of Blockchain (Shuaib et al., 2019, p. 3)	26
Figure 2.5.	An overview of Blockchain-based healthcare data management (Khezr et al., 2019, p. 5)	29
Figure 3.1.	Research roadmap	64
Figure 4.1.	Consumer interface for Microsoft HealthVault	81
Figure 4.2.	The validated E-Health ecosystem	85
Figure 4.3.	E-health digital flows: A simplified model	88
Figure 4.4.	A concept map of findings about the e-health ecosystem	102
Figure 5.1.	Singapore's distribution of healthcare services	105
Figure 5.2.	Singapore's Regional Health Systems network	107
Figure 5.3.	A concept map of Singapore's e-health ecosystem	114
Figure 6.1.	Number of US hospitals by ownership type (Total: 5627)	121
Figure 6.2.	Community hospitals by ownership type (Total: 4926)	122
Figure 6.3.	A concept map of the US e-health ecosystem	132
Figure 7.1.	Unique and common themes from the Singapore and US case studies	142
Figure 7.2.	Payoff graph for healthcare provider strategies	144
Figure 7.3.	Payoff matrix for e-health adoption game	146
Figure 7.4.	Stages of connected health	148
Figure A.1.	E-health ecosystem	163

Glossary of Abbreviations

ACA	Affordable Care Act / ObamaCare
ACO	Accountable Care Organizations
AIC	Agency for Integrated Care
ARRA	American Recovery and Reinvestment Act
B2B	Business-to-Business
B2C	Business-to-Commerce
B2G	Business-to-Government
BI	Business Information System
C2C	Consumer-to-Consumer
CAGR	Compound Annual Growth Rate
CCDA	Continuum of Care Document Architecture
CCHIT	Certification Commission for Healthcare Information Technology
CDS	Clinical Decision Support
CHIP	Children's Health Insurance Program
CLEO	Clinic Electronic Medical Record and Operation System
CPM	Cost Per Mille
CPOE	Computerized Physician Order Entry
eHINTS	Electronic Health Intelligence System
EHR	Electronic Health Records
EMR	Electronic Medical Records
EMRX	Electronic Medical Record Exchange
FDIC	Federal Deposit Insurance Corporation
GDP	Gross Domestic Product
GP	General Practitioner
HIE	Health Information Exchange
HIMSS	Healthcare Information Management Systems & Society

HIPAA	Health Insurance Portability and Accountability Act
HIT	Health Information Technology
HITECH	Health Information Technology for Economic and Clinical Health
HMO	Health Maintenance Organizations
HSOR	Health Services & Outcomes Research
IaaS	Infrastructure as a Service
IDS	Integrated Delivery Systems
IT	Information Technology
LOINC	Logical Observation Identifiers Names and Codes
MCO	Managed Care Organizations
MOH	Ministry of Health
NCPDP	National Council for Prescription Drug Programs
NEHR	National Electronic Health Record
NGNBN	Next Generation Nationwide Broadband Network
NHG	National Healthcare Group
NRF	National Research Foundation
NTU	Nanyang Technological University
ONC or ONCHIT	Office of the National Coordinator for Health Information Technology
P4P	Pay for Performance
PaaS	Platform as a Service
PDPA	Personal Data Protection Act
PHR	Personal Health Record
POS	Point of Service
PPO	Preferred Provider Organizations
RHS	Regional Health Systems
RHIO	Regional Health Information Organizations
ROI	Return on Investment
SaaS	Software as a Service
SIGIDE	Special Interest Group on Interactive Digital Enterprise
SingHealth	Singapore Health Services Corporation
SNOMED	The Systematized Nomenclature of Medicine
TCM	Traditional Chinese Medicine
TDHS	Total Digital Health System
UPMC	University of Pittsburgh Medical Centre

Contents

Preface	v
Overview of the Book	vii
List of Tables	ix
List of Figures	xi
Glossary of Abbreviations	xiii

1. Introduction to Digital Health — **1**

1.1	Nature of the Research Problem	1
1.2	Prior Research and Gaps	4
1.3	Research Questions, Scope and Design	6
1.4	Significance of the Research	10
1.5	Organization of the Monograph	10

2. An Overview of Emerging Digital Health Informatics* — **12**

2.1	Data Connectivity	12
2.2	Wearable Sensors	21
2.3	Data Storage and Security	25
2.4	Data Analytics and Modelling	36
2.5	Synthesis and Key Take-Aways	47
2.6	Chapter Recap	51

3. Review of Key Concepts and Methodology — **52**

3.1	E-Health as Defined in Extant Research	52
3.2	Major Barriers to E-Health: A Game Theoretic View	53
3.3	Vocabulary of Research Concepts	59
3.4	Selection of Cases for Phase 2	60
3.5	Choice of Research Methodology: Rationale	63

3.6	The Research Approach	63
3.7	Phase 1: Development of Conceptual Model of a Patient-Centric E-Health Ecosystem	64
3.8	Phase 2: Validation of Conceptual Model in the Field	70
3.9	Phase 3: Cross-Case Analysis	76
3.10	Chapter Summary and Recap	77

4. Conceptual Model of E-Health Eco-System — 78

4.1	Data, Analysis and Discussion	78
4.2	Caveats	101
4.3	Chapter Summary and Recap	101

5. Within-Case Analysis of Singapore's NEHR — 104

5.1	Introduction	104
5.2	Singapore's Healthcare Transformation Journey	105
5.3	NEHR Implementation	108
5.4	Singapore's E-Health Ecosystem	113
5.5	Lessons Learnt	117
5.6	Conclusions	118
5.7	Chapter Summary and Recap	119

6. Within-Case Analysis of the US' Hitech — 120

6.1	Introduction	120
6.2	The US's Healthcare Transformation Journey	123
6.3	HITECH Act Implementation	124
6.4	The US' E-Health Ecosystem	132
6.5	Lessons Learnt	136
6.6	Synthesis of Case Findings	137
6.7	Chapter Summary and Recap	138

7. Cross-case Analysis and Findings — 139

7.1	E-Health in Singapore and the United States	139
7.2	Health Statistics: Singapore vs the United States	140
7.3	Similarities and Differences across Cases	141
7.4	Optimal Ecosystem: A Game Theoretic View	143
7.5	Emergent Themes and Patterns	146
7.6	Sustainable E-Health Ecosystem: Critical Success Factors	150
7.7	Chapter Summary and Recap	153

8. Discussion and Conclusions 154

8.1 Summary of Key Findings: Answers to Research Questions 154
8.2 Implications for Researchers and Practitioners 157
8.3 Contributions and Limitations of the Research 158
8.4 Suggestions for Further Research 159

Appendix A. Expert Interview Template 161
Appendix B. Expert Interviews Coding Scheme 167
Appendix C. Case Study Interview Template 182
Appendix D. Case Study Interviews Coding Scheme 183

References 199

Index 237

CHAPTER

1

Introduction to Digital Health

1.1 Nature of the Research Problem

It is no exaggeration to say that an "e-revolution" has radically transformed the conventional landscape of business and consumerism, as is evident from the variety of e-initiatives successfully launched over the past several decades. Today, with the Internet and digital services dominating most key aspects of day-to-day living, online means of communication, entertainment, education, banking and a host of e-commerce transactions are not merely convenient but tangibly efficient, cost-effective and time-saving. However, despite the Internet having revolutionized most aspects of human life, its foray into healthcare has been relatively inconsequential possibly on account of the complexities inherent in the industry (Wickramasinghe, Fadlalla, Geisler & Schaffer, 2005; Hill & Powell, 2009; Black et al., 2011; Kellermann & Jones, 2013).

This is quite apparent, for instance, in the way the health data of the patient seeking medical help is typically collected and stored in the present day. It is common knowledge that such vital patient information continues to get recorded, processed and stored on paper in much the same way as done a century ago without recourse to information systems, decision aids and prompts (Middleton, 2008; Serbanati, Ricci, Mercurio & Vasilateanu, 2011; Jaroslawski & Saberwal, 2014). The danger involved in this situation is that the data stored on paper may get increasingly disjointed, incoherent or even inaccessible over a period of time as it gets passed on from hand to hand. This apart, inherent in this practice is the potential risk that any updates and changes incorporated on paper may not all come to the attention of the current healthcare provider which may, in turn, seriously compromise the quality of the healthcare provided.

The arrival of e-health may be recent, but its long-term potential is immense (Dyer & Thompson, 2001; Bulgiba, 2004; Tripathi, Delano, Lund & Rudolph, 2009; Parmar, Mackenzie, Cohn & Gann, 2014; Rothenhaus, 2015). A case in point is the dramatic rise in the number of people in the United States of America (US) looking for health information online, which jumped from 10 million in

2000 to 100 million in early 2007[1]. It was also estimated that by 2013 about 45% of US adults with a chronic condition would be using the Internet to manage their condition[2]. This trend, needless to say, is increasingly getting conspicuous all over the world including developing countries. This may be seen as an indication of healthcare consumers' "unquenchable need for more and greater access to health information and services" (Wen & Tan, 2003, p 2).

Coupled with the above observation is the policy imperative of many governments to reform healthcare by making substantial investments in health information technology (HIT) to improve its safety, quality and value (Clancy, Anderson & White, 2009; Black et al., 2011; Ross, Stevenson, Lau & Murray, 2015). Such trends are gaining traction globally, as evidenced by the plethora of e-health projects that have stemmed worldwide in recent years[3]. According to Markets and Markets (2015)[4], the world healthcare IT market is expected to grow to $228.7 billion in 2020 at a CAGR of 13.4% during the forecasted period of 2015 to 2020.

Ideally, e-health should encompass medical informatics, public health and business, and include within its purview health services and information that are delivered and enhanced through the Internet and related technologies (Eysenbach, 2001). With the integration of tele-health technologies with the Internet, e-health has the potential to enhance the quality and value of health services delivery through improved efficiencies and diminished costs thereby developing new markets (Wen & Tan, 2003; Baur, Fehr, Mayer, Pawlu & Schaudel, 2011). In essence, e-health comes with the promise of improved quality care, greater safety, reduced costs, reduced medical errors, increased efficiency of information flow and most importantly, empowerment of healthcare consumers in their healthcare decisions (Walker, Pan, Johnston, Adler-Milstein, Bates & Middleton, 2005; Vishwanath & Scamurra, 2007; Tripathi et al, 2009; Ebel, George, Larsen, Neal, Shah & Shi, 2012).

Although e-health envisages endless possibilities for the healthcare industry, there are several barriers to its adoption, the most significant ones being those that come from the healthcare providers' perspective. Some of the major barriers are high investment costs (Reed, 2007), uncertain returns on investment (Steele, 2006), loss of productivity (Clarke & Meiris, 2006) and, most significantly, misalignment of incentives (Glaser, 2007). Overcoming these deterrents is crucial because healthcare providers are supposedly the harbingers of the future

---

[1] Manhattan Research LLC Survey, 2007

[2] Manhattan Research LLC Survey, 2013

[3] https://www.moh.gov.sg/content/dam/moh_web/Publications/Information%20 Papers/2014/NEHR/English%20Brochure%20(Final).jpg; http://www.computerweekly. com/news/2240215175/UK-shows-biggest-take-up-of-electronic-Health-records-in-Europe; Arizona telemedicine network, USA

[4] http://www.marketsandmarkets.com/Market-Reports/healthcare-it-252.html

of e-health. If the barriers faced by them go unresolved, their participation and cooperation cannot be secured. As a result, the ideal of a patient-centric e-health system may not materialise.

Currently, healthcare providers may feel that they are unduly burdened with the responsibility of promoting e-health through investments in Electronic Health Records (EHR) systems. This may be because building such a system would obviously require huge investments and maintenance costs of equal magnitude. Further, costs may also come in the form of licensing and upgrading fees from time to time. Despite such investments, there is no guarantee of returns, however, owing to a lack of demonstrable evidence for the long term sustainability of an e-health system (productivity paradox). While the investment in EHRs is considerable - not only in terms of direct costs but also in terms of the time spent on staff training and the consequent loss of productivity, the returns on such investments may often be disappointingly low to warrant any justification.

To create a patient-centric e-health network, the combined power of technology and the Internet must be harnessed to foster a totally 'connected' health network that encompasses all the key stakeholders, and provides a common platform for interfaces and transactions among them, seamlessly connecting them in the process, for an exchange and reuse of health information. Since such a network is in fact an interconnected 'network of networks' that delivers a product or service through both competition and cooperation, it can be thought of as a 'business ecosystem'. James Moore, who pioneered the concept in 1996, describes the ecosystem as being made up of "customers, market intermediaries (including agents and channels, and those who sell complementary products and services), suppliers, and of course, oneself" (Kandiah & Gossain, 1998, p. 29). In addition, such an ecosystem should be able to create value for its customers by providing additional information, goods, and services, through the use of the Internet and related technologies (Kandiah & Gossain, 1998). The type of patient-centric e-health network envisaged in this monograph may be said to have the attributes that characterize a business ecosystem and may henceforth be referred to as an "e-health ecosystem".

Such an ecosystem may however be a difficult proposition in the current lopsided scenario where one stakeholder in particular, namely the healthcare provider, views itself as creating more value than it can capture from the network, with the other stakeholders benefiting more from the value created, a phenomenon known as the "**tragedy of the commons**".

A patient-centric e-health network is also expected to reduce information gaps in the provider-patient relationship, benefitting patients and empowering them in their healthcare decisions and choices. Such an outcome may not be desirable for healthcare providers who have been traditionally leveraging this information gap (**information asymmetry**) to their advantage. And given the huge investments they need to make in order to progress into e-health, they may feel that it is neither logical nor reasonable to have to share the benefits of their investments with others including patients.

Even though some healthcare providers may seem willing to share their patients' health data over the network, they may only want to do so within a private network. A private network is an arrangement entered into by players (strategic decision makers) for mutual benefits. Data is strictly shareable only within the network thus restricting patients' healthcare choices to such players as are part of the network **(information blocking)**.

Issues (dilemmas) such as the ones discussed above have not, as yet, been addressed and resolved to the satisfaction of healthcare providers. In this context, research has an important role to play inasmuch as it can establish the fact that a patient-centric e-health network may be feasible provided that certain conditions are met. To resolve these issues, appropriate trade-offs between conflicting notions such as fairness and efficiency must be achieved for every key player in the e-health network, particularly the healthcare provider. Fairness in this context would mean the pay-off received by a player proportionate to its contribution to the achievement of the total output, whereas efficiency would mean the benefits resulting from reduced information asymmetries.

1.2 Prior Research and Gaps

Over the last decade and a half there have no doubt been several studies exploring the potential opportunities of e-health and its resultant benefits to the key players, but these studies were, by and large, limited in scope and findings, perhaps owing to the nascence of the field.

For example, Parente (2000) and Wen & Tan (2003) based their study on the business models of the then-existing health e-commerce websites, Aggrawal & Travers (2001) highlighting some innovative changes that could be introduced into healthcare through B2B and B2C e-commerce business models and Joslyn (2001) showed the significance of patient-centric e-health business models in the context of rising consumerism. While these studies have made a meaningful contribution inasmuch as they helped identify some of the key players in the field as well as recognizing new values that are likely to be created and captured in the e-health network, they were, as a rule, narrowly focused in that they failed to take a holistic view of the e-health ecosystem, or adequately representing all the key stakeholders or players in terms of their roles and interactions.

Later, deBrantes, Emery, Overhage, Glaser & Marchibroda (2007) explored the potential of health information exchanges (HIEs) to function as economically sustainable intermediaries that could create value by reducing the information asymmetries among its customers, namely the healthcare market players, through information feedback loops. The study specifically focused on the values generated for the e-health market players through such information feedback loops, but barely touched upon business model arrangements which are necessarily a part of such systems.

Busch (2008) identified some key market players in the healthcare continuum and classified them as primary and secondary depending on how the players used health information – whether for direct and indirect patient-care related activities or for roles outside of these patient-care activities. Even though the roles of these players were clearly mapped, Busch's notion of value was from an audit perspective; it largely dealt with how a health information system should be audited for content, infrastructure and process to ensure appropriate internal controls, and not so much with how to organize the e-health ecosystem.

Raghupathi & Kesh (2009), on the other hand, examined the concept of total digital health systems (TDHS) that could offer both intra- and inter-enterprise benefits by fostering a sharing of health information among the various healthcare delivery participants. However, the focus of the study was on the TDHS technical design issues rather than on the design of a business model to organize the players in the e-health ecosystem.

DesRoches et al. (2010) conducted a study of the US hospitals to determine the relationship between EHR adoption and key metrics like quality and efficiency, and found a strikingly weak relationship between them. This led them to acknowledge the lack of evidence on how best to implement the EHR to achieve maximum gains in healthcare. Mensink and Birrer (2010) discussed the case of the Dutch Electronic Health Record, the progress of which they found to be slow, one of the reasons being the strategic considerations of the various players involved.

Paun et al. (2011) recognized the significance of both local and global interoperability of EHRs, as health data must be accessible to patients anytime and anywhere, and advocated an openEHR. The scope of their study was limited to a technical modelling of the openEHR for the Romanian healthcare system, without much discussion on business model arrangements within the ecosystem.

Wiedemann (2012) once again broached the topic of lack of evidence in favor of EHRs, and attributed that to unintended consequences of implementing and using EHRs that might create new risks and threaten patient safety. In conclusion, she raised the question whether or not a central agency would be needed to monitor the EHRs so as to prevent any negative consequences of EHR implementation and use by healthcare organizations. Here again, the focus was confined to EHR implementation and use-related issues within a healthcare provider setting.

Kellermann and Jones (2013) found the performance of Health IT in the United States 'disappointing' with reality being a far cry from what was projected and hoped for; seven years after RAND Corporation's projection in 2005 that $81 billion annual savings could be achieved with Health IT adoption, the annual healthcare expenditure in the United States actually grew by $800 billion. Some technical reasons suggested as being responsible for this phenomenon were sluggish health IT adoption, adoption of non-interoperable systems and failure to re-engineer care processes.

Rudin, Jones, Shekelle, Hillestad & Keeler (2014) observed that stakeholders in healthcare organizations were well aware of the benefits they could derive by being part of a health IT platform that would facilitate an exchange of health information among them. However, they lacked clarity in terms of how to organize this platform to create a compelling business case for its sustainability.

As recently as 2015, Bergmo (2015) highlighted the lack of a viable business model to deliver and sustain e-health. The author lamented the fact that in spite of e-health having been around for many years, "basic issues" with regard to its sustainability stand unresolved.

With e-health gaining momentum around the world, there are bound to be changes to the scope of traditional roles and processes in the e-health ecosystem, which may have a significant impact on e-health business models. However, it is evident from the above literature review spanning the period between 2000 and 2015, that little research is available on the potential critical success factors for a patient-centric e-health business model that considers the key players, and the values created and captured by each of them all at once, while suggesting alongside a win-win arrangement among them for a fair, efficient and sustainable e-health ecosystem. It is hoped that the present monograph will fill some of the gaps in the e-health literature and make a worthy contribution to this body of knowledge.

1.3 Research Questions, Scope and Design

Following the research gaps identified, this monograph sets out to investigate two fundamental research questions:

Research Question 1 (RQ1): *Who are the key players in a patient-centric e-health ecosystem and what are the potential values they can create and capture through digital data flows?*

The objective of RQ1 is to adequately model a patient-centric e-health ecosystem in terms of the key players, their roles, and the potentially major digital flows among them. Such modelling, it is hoped, would be a prerequisite to the identification of values created and captured in the e-health ecosystem for any subsequent analysis. RQ1 considers the case of healthcare providers in particular who are fraught with dilemmas of participation and cooperation in e-health. This is addressed in the first phase of this research.

Research Question 2 (RQ2): *What are the critical success factors for developing a sustainable patient-centric e-health ecosystem?*

RQ2 will help identify some key factors that have the potential to drive healthcare providers' decisions to invest in e-health, and share the benefits of their investments with other players. RQ2 is also intended to contribute to factors that need to be considered from healthcare providers' perspective in the design of a business model; an important consideration in the third phase of the research.

Healthcare providers may be the most important drivers of, and contributors to e-health, but it is wrong to assume that a patient-centric e-health system can be

Introduction to Digital Health

arrived at just by resolving their dilemmas. Therefore, the purpose of RQ2 is also to investigate if there is sufficient incentive for all the key players to participate in a patient-centric e-health ecosystem, by analyzing the values they create and capture. Although the word "sufficient" is subjective, it is not unreasonable to conjecture that the key players will not be motivated to participate in e-health if the values they capture (payoff) do not justify the values they create (contribution) in the network. On the other hand, without the participation of all the key players, e-health may not have the potential to generate sufficient values for its sustainability.

The primary objective of RQ2 is thus to identify the critical success factors for a sustainable e-health business model iteratively, using economic notions of fairness and efficiency for all players in a patient-centric e-health ecosystem. RQ2 is addressed in phases 2 and 3 of this research.

The scope of e-health considered for this research is a total digital health system (TDHS) connecting all the key players electronically via the Internet. In the actual sense though, e-health is a term that encompasses much more than just the Internet and healthcare. It is not just a technical development, but a "state-of-mind, a way of thinking, an attitude, and a commitment for networked, global thinking, to improve healthcare locally, regionally, and worldwide by using information and communication technology" (Eysenbach, 2001, p 20).

Figure 1.1 provides a high-level overview of the research methodology adopted in this monograph.

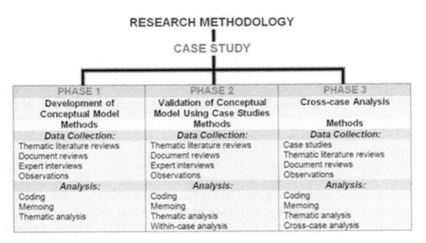

Figure 1.1. Research design

The appropriateness of a research method derives from the nature of the social phenomenon to be investigated. Given the nature and scope of the Research Questions, the most appropriate research methodology chosen for research purposes is the case study. Patton (1987) holds the view that if the phenomenon in question is an area of interest that is complex and calls for in-depth probing, the most appropriate method is the case study. Leedy & Ormrod (2010) concur that case studies are particularly apt for a researcher seeking to explore a little known or understood situation in depth. Given the multifaceted, complex and still-evolving nature of e-health, and the phenomenon under study, there is a good fit between the study and the chosen methodology. What makes e-health a complex phenomenon is that it involves several interwoven issues and externalities that need to be contended with in order to gain a holistic understanding of the phenomenon. Compounding this challenge is the fact that the literature concerning the evolution of the phenomenon is narrowly-focused as already discussed in section 1.2, 'Prior Research and Gaps'. The forenamed characteristics of this study make it an ideal candidate for the case study methodology which is believed to bring to the fore meaningful emergent concepts to help unravel conceptual relevance and ultimately lead to answers to the research questions.

Correspondingly, the research conducted in this monograph adheres to the following three phases:

1.3.1 Phase 1 – Development of a Conceptual Model of Patient-Centric E-Health Ecosystem

This phase of the study began with a systematic review of extant literature on health information technology (HIT) and observations of developments in the field to generate sensitizing concepts which would lead to the identification of the areas of enquiry for this research, commonly referred to as 'substantive' areas' – major streams of research explaining the topic. The substantive areas in this research were defined as 'e-health', 'game theory' and 'business model'. A focused and extensive review of literature and other information sources related to e-health then followed with the purpose of gathering data. This would help identify the key players in the e-health ecosystem and understand their roles and business model arrangements to adequately model the ecosystem as well as the major digital flows among the key players in it. The six primary or key players identified were - Patients (Consumers), Providers, Payers, Vendors, Infomediaries and Regulators.

In addition, key business model design parameters for each of the six identified key e-health players were analyzed using the ADVISOR business model framework, to make an assessment of the values they can possibly create in the network as well as to understand how and why they may collaborate for value rather than compete for revenues. A qualitative analysis of the e-health network supported by literature was then undertaken to identify the significant values created in e-health by each of the key players, and the corresponding values captured by them. These values have been presented in Table 4.4 in Chapter 4. A provisional model of the e-health ecosystem was thus developed.

Further, to validate the provisional model, inputs were sought from experts from different walks of the healthcare industry using an interview template developed for the purpose. These expert interviews were conducted through

Introduction to Digital Health   9

email, phone or personal meetings. Twenty-one industry experts were targeted for this exercise and the response rate was 57% with twelve experts completing the interview by the September 30, 2010 deadline. The provisional model was refined by comparing it with the data gathered through expert interviews, and perusal of evolving literature on the subject with the result a conceptual model for this research was educed through triangulation of the data sources.

1.3.2 Phase 2 – Validation of Conceptual Model Using Case Studies

Phase 2 of this research involved a descriptive case study approach to validate the conceptual model where the 'case' or 'unit of analysis' was a national level e-health ecosystem. This phase commenced with a review of emerging digital health technologies and the various healthcare systems around the world, which are presented in Chapters 2 and 3 respectively. The review, aided by literature reviews and tracking of industry trends, led to the identification of potential candidates for the case study. Representatives of these ecosystems were then identified through referrals, Internet searches and LinkedIn lookups, and contact was established with them via email to request their participation in the study. The criteria established for selecting cases were progress and performance. The above measures resulted in the selection of specific case studies to be conducted to seek answers for RQ2. The cases eventually selected were the national-level e-health ecosystems that are underway in Singapore (a high-performing healthcare system) and in the US (a low-performing healthcare system). The case study protocol was then developed, and case studies of the National Electronic Health Records (NEHR) system in Singapore and the Health Information Technology for Economic and Clinical Health (HITECH) program in the US, were conducted. The sources of data for the case studies included an environmental scan of the case or ecosystem through literature and documentation reviews, interviews with representatives of key players in the ecosystem and observations of characteristics of the ecosystem. The ecosystems were individually analyzed using the game theoretic notions of fairness and efficiency, and modeled through iterative conceptualization until triangulation of data sources was achieved. With-in case reports were prepared for the NEHR and the HITECH program.

Incidentally, the interview respondents are referred to as 'informants' in this study as they were selected through the 'key informant technique', where "one or a few individuals are solicited to act as guides" (Tongco, 2007) to the community of interest, which in this context was a national-level e-health ecosystem. Prerequisites for such informants were knowledge about the e-health ecosystem they are a part of, capacity to relate to the researcher's frame of reference and of course, willingness to participate in the study.

1.3.3 Phase 3 – Cross-Case Analysis

This final phase of the research focused on a cross-case analysis of Singapore's NEHR and the US' HITECH Program to uncover any patterns for the purpose of analytical generalization, leading to some modification of the original conceptual model. The modified model was compared with literature until theoretical saturation was reached, and a cross-case report prepared. Finally, a theoretical framework for conceptualizing the issues surrounding e-health adoption by key players, and identifying the critical success factors for a patient-centric, sustainable e-health ecosystem, was developed.

1.4 Significance of the Research

E-health has the potential to transform the healthcare industry that is currently fraught with problems of accessibility, quality and affordability. There is reason to believe that EHR is a key component of e-health. Notwithstanding the promises it brings, e-health, like any other high-technology environment, is faced with formidable legal, economic and operational barriers to its adoption. These typical barriers apart, a significant deterrent to patient-centric e-health is the adoption and use of EHRs by healthcare providers who tend to shirk from the investment because of the participation and cooperation dilemmas detailed in section 1.1, 'Nature of the Research Problem'. For the feasibility of patient-centric e-health, it is critical that these dilemmas be resolved to the satisfaction of the healthcare providers. As for its sustainability, it is critical that payoffs correspond to contributions for all key players, especially, the healthcare providers.

The findings of this monograph would hence make a meaningful contribution to the body of knowledge on e-health in terms of facilitating a better understanding of the e-health ecosystem, its key players and their roles and business model arrangements. Besides, the findings would also throw light on the values they can create (contribute to) and capture from (payoff) the ecosystem, and whether the apportionment of such values is equitable for all players. However, the most significant contribution of this study would be to generate a set of critical success factors for a sustainable patient-centric e-health business model that may eventually pave the way for 'collaboration for value' rather than 'competition for dollars'.

1.5 Organization of the Monograph

This monograph comprises eight chapters followed by references and appendices. Following this Introduction in Chapter 1, the remainder of this monograph is organized into the following chapters:

1.5.1 Chapter 2: An Overview of Emerging Digital Health Informatics

Chapter 2 discusses current trends in digital health technologies that demonstrate promise in enabling personalized digital healthcare. These technologies are grouped and reviewed under three categories namely: (1) Data Connectivity, (2) Data Storage and Security and (3) Data Modelling and Analytics.

Introduction to Digital Health 11

1.5.2   Chapter 3: Review of Key Concepts and Methodology

This chapter provides a systematic review of relevant literature which helped develop sensitivity to the phenomenon under study namely, e-health. Furthermore, the chapter details the steps followed in executing the three phases of this study namely: (1) Development of a conceptual model of the e-health ecosystem, (2) Validation of the conceptual model using case studies and (3) Cross-case analysis.

1.5.3   Chapter 4: A Conceptual Model of the E-Health Ecosystem

In this chapter, a conceptual model of the e-health ecosystem is developed in response to RQ1, using data sources such as literature and document reviews, expert interviews and observations.

1.5.4   Chapter 5: Validation of Conceptual Model: Within-Case Analysis of Singapore's NEHR

Chapter 5 covers the first of the two case studies undertaken for the purpose of validating the conceptual model. The case in question is the NEHR, an e-health initiative led by the Singapore government. A within-case analysis of the NEHR is presented as part of this chapter.

1.5.5   Chapter 6: Validation of Conceptual Model: Within-Case Analysis of The US' HITECH Program

Chapter 6 presents the within-case analysis of the HITECH Program, an e-health initiative spearheaded by the US government. This forms the second case study conducted to validate the conceptual model.

1.5.6   Chapter 7: Cross-Case Analysis

This chapter addresses Phase 3 of this study. A comparison is made between the two case studies - the NEHR and the HITECH program, to produce cumulative knowledge and seek a common explanation that can characterize the issues surrounding e-health implementations.

1.5.7   Chapter 8: Conclusions

Chapter 8 discusses the major conclusions from this research - the critical success factors for e-health implementations. It furthermore highlights the contributions of this study to the existing body of knowledge and describes the study's limitations as well. The chapter concludes with suggestions for further research.

The next chapter, Chapter 2, provides an overview of emerging digital health technologies that represent great potential to facilitate innovative personalized healthcare services through e-health platforms.

CHAPTER

2

An Overview of Emerging Digital Health Informatics*

The application of digital technologies to healthcare, while not nascent, remains a rapidly evolving area of research and practice. Innovative developments in digital health have spanned the network, storage and computational aspects of the eco-system. This chapter provides an *ex-post* overview of emerging information technologies for digital health. It comprises detailed descriptions and up-to-date research on the emerging information technologies that impact the future of digital healthcare. The chapter is organized into sections as follows: (1) Data Connectivity, (2) Data Storage and Security, (3) Data Modelling and Analytics, (4) Synthesis, (5) Recap. The chapter concludes that, given current trends in digital technologies, personalized digital healthcare is the ultimate and inevitable outcome of these developments.

Figure 2.1 shows the topics of emerging information technologies for digital health described in this chapter. The framework is helpful in specifying the scope and coverage of the chapter as well as identifying any gaps in the literature.

2.1 Data Connectivity

This category includes information technologies such as the Internet of Things (IoT) and Wearables, which are commonly used for connectivity and data transfer. Discussion includes definitions, brief history, working mechanisms, implications in healthcare as well as findings from recent studies on the topic.

2.1.1 Internet of Things (IoT)

The Internet of Things (IoT) is network-enabled technologies like mobile devices and wearable sensors, which are able to sense, actuate and interact with other related devices through the Internet (Sheth et al., 2018). Since its inception,

---

* An invited contribution from Arnob Zahid. The author would like to acknowledge ICT Division, Ministry of Posts, Telecommunication and Information Technology, Bangladesh for supporting this research.

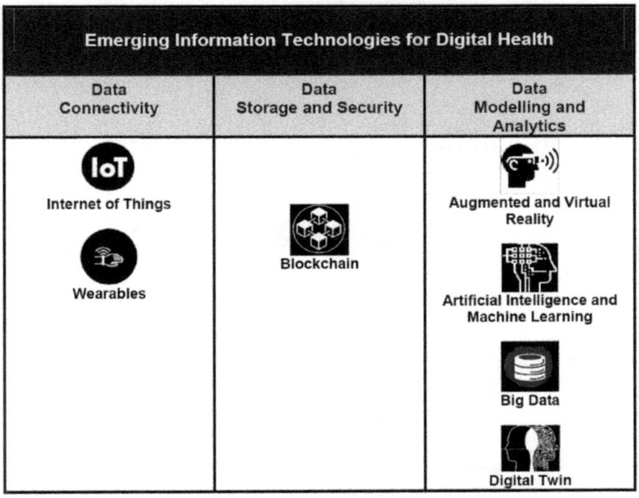

Emerging Information Technologies for Digital Health		
Data Connectivity	Data Storage and Security	Data Modelling and Analytics
IoT Internet of Things Wearables	Blockchain	Augmented and Virtual Reality Artificial Intelligence and Machine Learning Big Data Digital Twin

Figure 2.1. Chapter framework of emerging information technologies for digital health.

IoT is redefining trends in information creation, ingestion and sharing (Sheth et al., 2018). In the beginning, IoT was envisioned to extend the term "Internet" into physical objects of daily life using Radio Frequency Identification (RFID) technology (Acampora et al., 2013; Naranjo-Hernández et al., 2012). But lately, due to the rapid advancement of sensing technologies (development of various heterogeneous sensors like accelerometer, altimeter, gyroscope) and other low-cost portable devices, its horizon has grown extensively (Qi et al., 2017).

IoT enabled personalized healthcare generally consists of four layers (Figure 2.2): (a) Sensing Layer, (b) Networking Layer, (c) Data Processing Layer, and (d) Application Layer. The Sensing Layer is assigned to monitor patients' overall health condition (physical, mental and emotional) (Wan et al., 2018). To perform the observation, a set of embedded sensors can be applied (C. Lin et al., 2009). This may include wearable sensors for ECG and blood pressure (to accumulate the biomedical parameters), RFID for identification, and GPS for localization and positioning (J. Wan et al., 2018). Smart homes are also in use (often) for sensing the immediate surroundings (e.g. home conditions, items used) of the patient (Shu et al., 2010).

The Network Layer ensures secure, efficient data transmission to the corresponding data processing unit (Guo et al., 2011). To perform data transmission, several short-range communication protocols are widely applied (for example, Zigbee) (J. Wan et al., 2018). Several new techniques like NB-IoT (Sinha et al., 2017), LoRa and 6LowPAN (Qiu & Ma, 2018) have also been introduced in recent times. The Data Processing Layer is responsible for extracting valuable knowledge derived from the Sensing Layer. Here, the learning-based approach can be applied to ensure effective data mining. Lastly, based on the top three layers, the Application Layer provides different intelligent services and

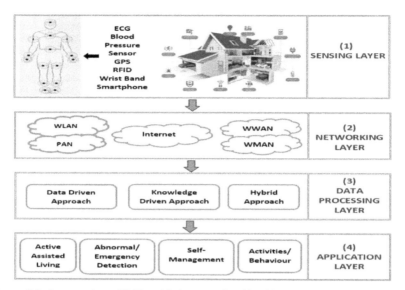

Figure 2.2. An overview of IoT enabled personalized healthcare (Wan et al., 2018, p. 2).

applications like disease diagnosis, smart assistance and behavior recognition (Jie Wan et al., 2018).

In healthcare, IoT is playing a vital role in developing of healthcare and medical information systems (Chiuchisan et al., 2014). The technology is increasingly integrated into different healthcare equipment (e.g. body scanners, heart monitor and wearables) thus enabling those devices to collect, process and share health data through the Internet in real-time. Using IoT, people are now keeping track of and tracing their food habits, sleep, vital signs, physiological status and other relevant activities (Sheth et al., 2018). Healthcare providers also use IoT to capture images, detect malignancy or suspicious cells and share the acquired knowledge with rightful individuals such as patients (Khezr et al., 2019). Figure 2.3 illustrates a generic overview of the IoT paradigm in healthcare, augmented by other emerging technologies like Blockchain, smart contracts, artificial intelligence and cloud computing.

- Step 1: Patient; the main source of all data.
- Step 2: Attached IoT devices for close or remote monitoring of the patient. Generates a large volume of data.
- Step 3: Data from step 2 is stored either in blocks or in cloud storage. AI helps Blockchain to generate intelligent virtual agents for creating new ledgers automatically. To protect the sensitive data of the patient, decentralized AI systems help Blockchain provide higher security ("Decentralized AI: How AI and Blockchain can work together," n.d.).
- Step 4: Healthcare providers (authorized by the patients); responsible for delivering safe and sound healthcare based on the patient data stored.

An Overview of Emerging Digital Health Informatics 15

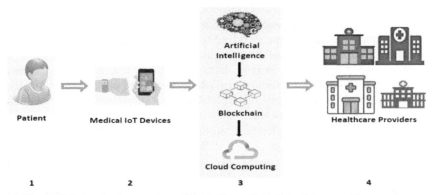

Figure 2.3. A generic overview of the IoT paradigm in healthcare with emerging technologies (Khezr et al., 2019, p. 15)

In personalized healthcare, IoT is still in its early stage (Qi et al., 2017). A lot of research has already been undertaken to harness its potential in healthcare delivery. The number of new research projects in the arena is growing at a significant pace. The majority of these research projects are now focused on personalized assistance for older adults with chronic conditions (e.g. diabetics, Alzheimer, heart disease) (Hallfors et al., 2018), to achieve independent living. For example, Doukas and Maglogiannis (2012) introduced a cloud computing-based pervasive healthcare platform which includes wearable sensors to collect patient data. The collected data is sent to remote cloud storage by smartphone (as an intermediary gateway), applying RESTful web interfaces where REST stands for Representational State Transfer, an architectural style for designing decentralized systems. To ensure secure data transmission, the platform maintains end to end security. However, the platform is unable to provide a high-level abstraction of the collected data for the patients, and does not use semantic web technologies (Datta et al., 2015). Romero et al. (2016) introduced a system to monitor and diagnose Parkinson's disease, whilst, Varatharajan et al. (2018) presented an early-stage Alzheimer disease detector using wearable IoT, based on a dynamic time warping (DTW) algorithm.

Apart from the above mentioned scholarly works, several research projects are also found in extant literature.

- Choudhury et al. (2008) designed a Mobile Sensing Platform (MSP) to recognize the various physical activities of a user. MSP is a lightweight wearable device (connected to a mobile phone) placed on the user's waist to detect various activities of daily living (ADLs). The device is acknowledged as "state of the art" in early stage activity monitoring and detection with wearable sensors. MSP includes three models (sensing model, feature processing model and classification model), in versions 1.0, 2.0 and 3.0. To measure different types of human activities, microphones and accelerometers are used in the device as distinct sensors. Version 1.0 of MSP had some issues with processing, storage and battery life, which were resolved in version 2.0. In versions 1.0 and 2.0, a supervised training approach was applied, and as a result, accuracy rates of 83.6% and 93.8% were achieved respectively. In version 3.0, labelled training data was reduced and a semi-supervised training method was approached to cluster the activity patterns automatically. This resulted in an accuracy rate of 87.4%. MSP trains with offline data but provides feedback in real-time.

- Storf et al. (2009) developed a multi-agent-based activity identification framework under the project EMERGE (Emergency Monitoring and Prevention) (*EMERGE Project*, 2007), which targets the emergency medical services (EMS) system to support elderly care. The framework is able to detect ADLs automatically in an IoT setting. For activity detection and measuring vital data, it collects data from wrist devices (including wearable sensors) and ambient sensors. The framework classifies various activities: short term emergencies (e.g. helplessness, fall) or long term medical assessment (e.g. sleep, usage of the toilet) by applying a knowledge-based approach and highlighting weight as a characteristic for determining fuzzy reasoning. It defines the relationship between facts with the sequence for temporal interference. The knowledge-based model is hierarchically constructed and used as an interference agent to describe objects and connection at the sensing layer. This framework has been applied in Europe by caregivers and older adults using a closed-loop principle (a system that uses feedback where a portion of the output signal is fed-back to the input to minimize errors and increase stability).

- Kwapisz et al. (2011) developed a system that is able to detect physical activities using sensors on Android-based phones in the user's pocket. The data is derived from an accelerometer in the system. Features are extracted in accordance with the identified time between signal peaks and activities (ascending and descending stairs, jogging, walking, standing and sitting). Activities are identified by their monotonous characteristics. A supervised training algorithm is used to identify and compare, including logical regression, straw man, multiplayer sensitivity and the J48 algorithm. The system's result shows that ascending and descending stairs is the most difficult one to identify. Moreover, the result extracted from the activities may vary due to the mobile phone placement on the user.

- SMART (Self-Management supported by Assistive, Rehabilitation and Telecare Technologies) project (Huang et al., 2011) provides a personalised monitoring and self-management platform to patients suffering from chronic health conditions (e.g. heart failure, stroke, chronic pain) using wearable sensors. The project is aimed at assisting patients at home. It helps patients maintain health conditions by setting life goals through numerous physical activities. It also provides feedback based on the activities' outcome and suggests therapy plans. To monitor patients' physical activities, it uses both:

An *Overview of Emerging Digital Health Informatics* 17

accelerometers (to monitor blood pressure, weight) and Access Activity Log (AAL) sensors (to monitor sleeping pattern, food preparation, television usage and other physical activities). SMART adopts both the knowledge-based method and the data-driven method.

- Banos et al. (2014) designed an open framework named mHealthDroid (Mobile Health Android) to encourage agile development of Android-based biomedical applications. mHealthDroid collects data by connecting with heterogeneous commercially available devices (smartphone, smartwatch, belt) for both biomedical signals and ambulation. The components of mHealthDroid include communication manager, data processing manager, storage manager, system manager and visualization manager. The data processing manager uses Weka (a data mining software) (Hall et al., 2009) to perform specialized activities such as data pre-processing, feature extraction, segmentation and classification. It also provides guidelines and alarms as a means of targeting health interventions. The most significant feature of mHealthDroid is its extensibility. For example, mDurance (a mobile support system for healthcare to assess trunk endurance) (Banos et al., 2015), enhances the core functionalities of mHealthDroid. Overall, it facilitates new system implementation with diverse models and working ways that save time and money.
- Qi et al. (2015) developed a knowledge-based decision support and evidential reasoning system which supports patients suffering from chronic diseases. The system was developed under the MOSKUS project (*Norwegian Computing Center*, 2015). It uses a set of reasoning rules to provide non-pharmacological care plans to patients. The system helps patients to keep control over their health and reduces the frequency of hospital visits. The self-reporting feature of the system (questionnaire based) is assessed through categorization (None, Low, Medium, and High). The system provides: information about health conditions, behavioral and medical assessment reports, and an inference mechanism (decision recommendation).
- Puppala et al. (2015) developed an integrated clinical informatics framework named METEOR (Methodist Environment for Translational Enhancement and Outcomes Research), which consists of a logic and data storage EDW (enterprise data warehouse), and a software intelligence and analytics (SIA) tool (to determine prediction on the clinical outcome) for use by caregivers, physicians and other clinical staff. METEOR is also designed to monitor (remotely) and control the patient's physical state, based on the data collected (through spirometry, temperature, blood pressure, pulse oximetry). Medical interventions and reminders are presented to the patient using a web browser. The entire engine integrates many vital techniques such as service-oriented architecture (SOA) and JBoss (an application server), where rules of reasoning are managed and extracted from the EHR. The framework is also used to monitor COPD (Chronic Obstructive Pulmonary Disease) patients, demonstrating its versatility.

To transform the conventional healthcare system into a personalized supported model, integration of wearable IoT (Dimitrov, 2016) with Blockchain (see Data Storage and Security section for detail) is a must (Talukder et al., 2018). Abundant research has been found in the literature addressing this need.

- Zhang et al., (2016) proposed a secure Pervasive Social Network (PSN) model for healthcare. The model's major challenge was ensuring security during data sharing among the PSN nodes. To resolve this issue, the researchers introduced two protocols. The first protocol was designed based on the enhanced version of IEEE802.15.6 and was capable of generating secure links with computational requirements for sensor nodes with limited resources and mobile devices. The other protocol was designed based on a Blockchain mechanism to ensure secure data sharing among the PSN nodes.
- To ensure the immutability of health records, Ichikawa et al. (2017) designed a Blockchain-based tamper-resistant mobile health system. The research was purposed to develop a smartphone application for providing cognitive behavioral therapy to people affected with insomnia.
- Jo et al. (2018) proposed a structural health monitoring system based on IoT and Blockchain. The proposed system activates both locally centralized and globally decentralized distribution by separating them into two different networks (edge and core network). It improves the efficiency and scalability of the Blockchain system.
- Rahman et al. (2018) proposed an intellectual dyslexia analytics solution, including a decentralized Big Data repository. The repository is used to store data before sharing it with healthcare groups, communities and individuals via Blockchain. The Mobile Multimedia Health data (generated during dyslexia testing) is stored in the decentralized Big Data repository and is sharable for conducting further statistical analysis and clinical research.
- Griggs et al. (2018) proposed a secure real-time patient monitoring and intervention system by integrating a Wireless Body Area Network (WBAN) with Blockchain smart contracts. The research proposed the integration of Blockchain to generate smart contracts, which will assist in evaluating information received by the patient's medical IoT devices. The evaluation occurs based on customized onset values. It resolves the issue of logging transfer, which usually appears during data transmission to an IoT based healthcare system.

The information and communication technology (ICT) infrastructure of today's healthcare system includes numerous networks, terminals, IoT, wearable sensors, wireless access points, and diagnostic tools. It facilitates various data transactions among patients, physicians and caregivers. The transacted data travels through various unknown communication networks, which may introduce the possibility of vulnerabilities (privacy or security breach) (Khezr et al., 2019). This issue has been addressed in various research papers.

An Overview of Emerging Digital Health Informatics **19**

- Nikoloudakis et al. (2019) introduced a software-defined network that can virtualize the conventional network structure into multiple abstraction levels to develop a resistance mechanism for defending malicious attacks. The authors also advised adopting a fog architecture (an architecture based on edge devices to execute a substantial amount of computing, storage, and communication over the Internet backbone), which can evaluate susceptibilities of existing and newly added hardware into healthcare IoT via OpenVAS (a software framework, consisting of several tools and services for vulnerability scanning and management). However, the proposed model was found inadequate to assess efficiently (the average time for a single evaluation is 7.21 minutes), impacting the system's scalability (Khezr et al., 2019). This latency issue can be resolved by adopting the Blockchain-based Ethereum platform, as it takes 10 to 15 seconds for block mining (Khezr et al., 2019).
- Nausheen and Begum (2018) proposed a different type of solution to secure key logic and mobile applications in other research. The solution is based on an application hardening approach (e.g. code complication) and API (Application Program Interface). The study advises adding a defense layer against reverse engineering. This may include checksum techniques and return-oriented programming. Overall, the research is found reliable in supporting the security and protection of medical devices and gadgets. Yet, heterogeneous technologies are also needed to achieve accumulative efficiency, bearing in mind the expense of increased complexity in the entire system (Khezr et al., 2019).
- To increase an Intrusion Detection System's (IDS's) capability in an IoT network, Otoum et al. (2018) proposed a technique called Adaptively Supervised and Clustered Hybrid (ASCH-IDS). The technique is able to aggregate sensory data to classify possible intruders. A disparate amount of aggregated sensory data is placed into two different detection subsystems for better performance: Misuse detection (MDSs) and Anomaly detection (ADSs). The system performs surprisingly well when an increase in the proportion adjustment occurs. It shows the highest accuracy in sensitivity and specificity when the proportion of misuse detection is found at 0:25% (indicates an increased detection rate).
- Otoum et al. (2019) deployed a Restricted Boltzmann Machine (deep learning-based methodology) Clustered IDS (RBC-IDS) to monitor critical infrastructure and detect potential intruders. The system uses three covert layers, which resulted in an increase in accuracy of 99.91% (in previous research, it was 99.80%). Both of these research projects are not cost-friendly, and expenses may rise with scale (Khezr et al., 2019).
- To resist attacks like DoS (Denial of Service), Remote to the user, Probe, Aloqaily et al. (2019) proposed an intrusion detection mechanism. The researchers came up with a deep belief network for data reduction, dimensions

and decision trees. They applied an ID3 algorithm for the classification of intrusion. The system's accuracy is very high (99.92%), though the false-negative rate is 1.53%. The result indicates that three intrusion attacks will be enough to penetrate the system per 200 user sessions (approximate) (Khezr et al., 2019).

- Saia (2018) and Saia et al. (2019) developed a novel approach for the internet of entities (IoE) by integrating mobile and IoT networks into two main modules: entities like user devices, and trackers which assist in information transmission to the Blockchain network. The developed architecture is an amalgamation of Blockchain and wireless networks which can be used in healthcare. However, implementing such a large and complex network may reveal prospective limitations. For example, the computational overhead optimization of Blockchain might be complex (Saia, 2018; Saia et al., 2019).

- Dorri et al. (2016; 2017) developed a lightweight novel architecture for a network supporting numerous IoT devices. In architecture, a local computer is utilized as a miner. Several IoT devices are clustered in such a way that one can be selected as a cluster head, using a policy header to decrease access verification. The scheme allows the policy header to keep an access control list (updated) which the network operator supervises. The proposed study utilizes a shared overlay network where numerous data owners can insert data in the blocks, and a single admin administrates the entire network. This may create mistrust among users (Khezr et al., 2019).

Besides data security, privacy and protection for IoT enabled personalized healthcare, interoperability is an important aspect that needs further attention. Numerous literatures have been developed addressing this issue.

- Brandt et al. (2013) addressed the interoperability issue utilizing semantic web technologies.

- Santos et al. (2015) developed an architecture that allows communication between personal health devices (PHDs) and the Internet. The architecture is based on SOAP (Simple Object Access Protocol) and comprises IEEE 11,073 standards. The authors also demonstrated that the proposed system is able to exchange data with legacy systems.

- Kasthurirathne et al. (2015) proposed a standardized API (domain-independent) for EMR systems. The authors adopted the modular architecture of OpenMRS (Open Medical Record System) to develop a FHIR (Fast Healthcare Interoperability Resource) based add on module, that is capable of dealing with requests placed on OpenMRS.

- Semenov et al. (2018) developed a patient decision aid system that is able to: collect data from resources that are available, analyze and read laboratory test results automatically, provide recommendations for precise diagnostics and ongoing tests, and deliver reports (automatically generated) to patients and physicians in a natural language. The system is based on production rules and adopts FHIR to provide semantic interoperability. The research findings offer

An Overview of Emerging Digital Health Informatics

a detailed understanding of the semantic interoperability of CDSS (Clinical Decision Support System).

- Cruz et al. (2018) proposed a HL7-OPC (Health Level Seven- OLE for Process Control) server, which allows data transmission among vital monitoring systems within PIMSs (Patient Information and Monitoring Systems) and allows storage of long historical vital sign data.
- Mavrogiorgou et al. (2019) proposed a mechanism to address IoT's interoperability issue in healthcare. The developed mechanism initially collects data from different connected devices and performs its cleaning. It then uses the cleaning results to achieve overall data quality levels for each collected dataset (in amalgamation with measurements from each devices' availability and consistency). Subsequently, only the data with high quality is stored and interpreted in a common format, which can be used for further purposes. The proposed mechanism was independently assessed in a specific setting and came up with reliable outcomes (data quality accuracy of 90% and interoperability accuracy of 100%).

2.2 Wearable Sensors

Just as they are named, wearable sensors simply mean wearable objects which are integrated with the human body (directly or indirectly) to monitor the health status of patients, and to provide health status data for ensuring appropriate healthcare delivery (Fang et al., 2017). In modern times, the application of wearable sensors to deliver patient-centric healthcare is growing at a rapid pace, along with other technological advancements. Wearable sensors nowadays are vastly integrated into IoT to deliver patient-centric healthcare. This has been thoroughly discussed in the IoT sub-section of this chapter. In the current market, various types of wearable sensors are available. To develop a clear understanding of this, Qi et al. (2017) classified wearable sensors into the following categories.

2.2.1 Inertial Sensors

These sensors are small-scaled MEMS (Micro-Electro-Mechanical System) devices, usually used to measure physical activity. Inertial sensors are positioned on different parts of the body (Veltink et al., 1996). This includes sensors like accelerometer, gyroscope, pressure sensor, and magnetic field sensors (Qi et al., 2017). The accelerometer is used to detect human motion based on measuring the degree of position change. Gyroscope is used to measure rotational movements like knee joint rehabilitation (Chen et al., 2015), combining an accelerometer in the measurement process. These sensors are used to detect specific types of human behavior and emotions with accuracy, which includes turning or ascending stairs (Salarian et al., 2007), descending stairs, and knee bending (Pappas et al., 2001). Other applications of these sensors include stroke rehabilitation (Reddy et al., 2010), gait rehabilitation, joint pathology (Lee et al., 2014), fall detection (Luštrek & Kaluža, 2009) and Parkinson's disease (Salarian et al., 2007). Pressure sensors combining accelerometers are also used to monitor fall detection (Bianchi et al., 2010) and stair behavior (Moncada-Torres et al., 2014), by studying their relationship with altitude and sensory readings. Magnetic field sensors are used to detect human direction with high spatial resolution by placing the sensor close to the measurement location (Qi et al., 2017). For example, an activity of "watching TV" can be specifically detected (including face direction) using a magnetic field sensor, combining an accelerometer (Lowe & Ólaighin, 2014).

2.2.2 Location Sensors

This only includes GPS (Global Position System), which provides location data (a default characteristic). GPS is able to provide a high level of information about the activities of a person. This includes a person's engagement with others or other activities over a period of time (day/week/month), and the relationship between the person's activities and location (Ashbrook & Starner, 2003; Liao et al., 2005; Liao et al., 2007). GPS is usually used to monitor location change activities (open air and/or mobile environment) (Ashbrook & Starner, 2003). GPS has applications in healthcare.

2.2.3 Physiological Sensors

These sensors are developed to measure specific health-related data (e.g. temperature, heart rate). Physiological sensors include Blood Pressure Cuff, Electrocardiogram (ECG), Spirometer, Electrooculography (EOG) and Galvanic Skin Response (GSR) (Qi et al., 2017). These sensors are mostly used in clinical settings and are expensive in comparison to others because of their higher accuracy. Due to technological advancements in sensory techniques, physiological sensors are becoming inexpensive (Qi et al., 2017). Electrocardiogram (ECG) for heart rate monitoring is contributing to patients' daily health monitoring (Giuffrida et al., 2008), physical activity detection and monitoring (Altini et al., 2015; Qi et al., 2015). These sensory devices are feasible to use in clinical settings by allowing data transmission via internet (Qi et al., 2017).

2.2.4 Image Sensors

In the IoT context, an image sensor is simply a camera, which is used to record human emotions, activities and contexts in image or video format. In IoT-enabled personalized healthcare (Chung & Liu, 2008) image sensors typically include cameras like Sony Xperia eye, and SenseCam. These low-cost wearable cameras are used to record human life activities (in sequence) as a visual life-logger. Based on image annotation and location data, these tools can effectively detect user behaviour and daily activities and result in enhanced and inventive home-based care solutions for older adults. However, compared to other sensor technologies, image sensors in IoT-based healthcare require high privacy and protection levels. The storage of video and image data in lifelogging mode is a big challenge, as data volume escalates with time (Qi et al., 2017).

An Overview of Emerging Digital Health Informatics **23**

In addition to the above wearable sensors, many commercial wearable devices are available in the market. These commercial devices are creating new opportunities for collecting multi-dimensional health data, derived from hybrid sensors. Examples include Fitbit Charge[1], Garmin Vivosmart[2], and Huawei Band[3]. These are all wristband devices, which record information about burnt calories, steps taken and distance covered by the user. The collected health data is synced to a mobile phone using Bluetooth and is used in related mobile applications.

Inertial sensors have made great progress in the last couple of years. However, they are constrained for long-term activity monitoring, especially in a free-living environment. Although it is a small sensor integrated on a certain part of the body, it gives an unpleasant feeling for users requiring permanent monitoring (Qi et al., 2017). Wearable devices have been very popular in recent times, although their usage is limited to fitness. These devices are manufactured to provide several processed measurements (e.g. calories, steps, distance), which comprise numerous uncertainties. Raw data from sensors can be acquired from a smartphone, but due to diversity in lifestyle and environmental effects, personal health data derived from wearable devices demonstrate notable ambiguity in a real-life context. Issues include capacity, battery life and placed position (Qi et al., 2017). In physical activity monitoring, the results differ when the smartphone is kept in the pant's pocket compared to in a bag. This is obvious as inertial sensors are very sensitive to position or orientation changes. Therefore, validation of mobile data from these sources for longitudinal healthcare is crucial (Qi et al., 2017) and warrants acute research focus.

Wearable sensors in IoT enabled personalized healthcare is still in an early stage. A detailed discussion on prior research of wearable sensors along with IoT has already been made in the previous section (IoT section) of this chapter. There is also other research in wearable sensors and assessment tools that promise significant contributions.

- Sardini et al. (2015) designed a wireless wearable T-shirt to monitor posture during reinforcement exercises or rehabilitation. The data obtained from the wearable sensor is found reliable, compared to data from optical systems.
- Wu et al. (2015) proposed a wearable biofeedback system using inconsistent heart rate to assist personalized emotional management. The proposed system was found to be considerably effective in sensing negative emotions.
- Billis et al. (2015) proposed a framework for independent living of older adults and were able to evaluate depressive symptoms (Geriatric depression case-study).
- Melillo et al. (2015) designed and developed a flexible, transparent, and extensible risk assessment system for hypertensive patients' vascular events.

---

[1]   Fitbit Charge[TM] Wireless Activity + Sleep Wristband (n.d.)

[2]   Fitness Tracker | Fitness Bands | Activity Tracker | Garmin (n.d.)

[3]   HUAWEI Band 3 Pro, Built-in GPS Sport Band, Smart Wearable/HUAWEI Global (n.d.)

The system uses data mining functionalities to deliver proactive remote monitoring.

- Pierleoni et al. (2016, 2015) developed a waist-worn wearable device using a biometric sensor and AHRS (Attitude and Heading Reference System) to detect older adults' falls. The developed device is found to be exceptionally efficient in detecting falls with 100% sensibility.
- Etemadi et al. (2016) developed a low-power multi-modal patch to measure a patient's activity using SCG and ECG sensors. The designed patch is able to measure environmental context, hemodynamic and combined activity, all with the same patch. The patch's operating power is more than 48 hours (with seamless recording).
- Thomas et al. (2016) developed a wristwatch-based blood pressure measurement system utilizing ECG and PPG.
- Xu et al. (2016) developed a contextual online learning method based on data from inertial sensors (inexpensive, body-worn) for classifying activity. The system is able to facilitate treatment in community rehabilitation and athlete coaching.
- Spanò et al. (2016) developed a remote ECG monitoring system focusing on non-technical users. The developed system is designed for long-term monitoring in a residential context and comprises broader IoT enablement.
- Chia et al. (2016) developed a smart ECG patch for measuring ECG utilizing three electrodes (integrated into the patch). The patch filters measured signals to decrease noise, determine analogue to digital conversion and senses R-peaks.
- Zhu et al. (2017) developed a fall detection system based on a passive RFID tag, using Doppler frequencies and RSS. The system's prototype resulted in 98% accuracy in behavior detection.
- Hooshmand et al. (2017) proposed a novel approach to reduce the power consumption of sensors in IoT enabled healthcare systems. The proposed approach allows a reduction in signal size up to 100 times and entails a similar reduction in energy demand, along with reconstruction errors limited to 4%.
- Khojasteh et al. (2018) developed a wrist wearable (accelerometer-based) fall detection solution at the optimization stage. The system is rule-based and incorporates similar neural network research. The system is cost-effective and supports vector machines (high specificity).
- Kheirkhahan et al. (2019) developed a real-time remote monitoring framework by assimilating a smartwatch-based application with a remotely connected server for concurrent data collection in EMA, activity, mobility, intervening health events and patient-reported outcomes. The framework provides real-time data visualization and summary statistics. The framework also complies with some requirements for next-generation IoT enabled healthcare.

2.3 Data Storage and Security

This category includes Blockchain and Smart Contracts, which are usually used for data storage and encrypted transactions (for security). Discussion includes definition, brief history, working mechanism, types and versions of Blockchain, benefits of using Blockchain, and implications in the healthcare sector and prior research.

2.3.1 Blockchain and Smart Contracts

Blockchain can be defined as a parallel and distributed computing mechanism along with a ledger, which is used to ensure trusted transactions over the Internet. The trust is recognized via peer to peer distribution networks and smart contract code, instead of the traditional centralized mechanism (used to authenticate transaction settlement) (Antonopoulos, 2017; Croman et al., 2016; Hardjono & Pentland, 2019; Kosba et al., 2016; Nakamoto, 2008; Poon & Dryja, n.d.; Vigna & Casey, 2016; Zhang et al., 2016). A Smart Contract is simply the code (inspired by the traditional pen and paper-based lawful contract), which is executed to impose the rules exactly how they are written (Shae & Tsai, 2018). When a transaction is made in a Blockchain network, it gets recorded on a distributed ledger which is neither changeable nor deniable (Shae & Tsai, 2017). Blockchain has been very popular as a Distributed Ledger Technology (DLT) since its inception in October 2008 (Nakamoto, 2008). Different industries around the globe are currently adopting or already have adopted Blockchain technology, including Finance (Beck et al., 2017; Cai, 2018; Zheng et al., 2018), the Energy sector (Andoni et al., 2019; Chitchyan & Murkin, 2018; Zhang et al., 2017), Privacy and Security (Conti et al., 2018; Joshi et al., 2018; Li et al., 2017; Lin & Liao, 2017), Government (Alketbi et al., 2018; Hou, 2017; Ølnes et al., 2017), and IoT (Banerjee et al., 2018; Conoscenti et al., 2016; Dorri et al., 2017; Fernández-Caramés & Fraga-Lamas, 2018; Ferrag et al., 2019). Besides these, Healthcare has also started adopting this technology for different healthcare applications (Agbo et al., 2019; Shuaib et al., 2019), including PHC (Hussein et al., 2018; Juneja & Marefat, 2018; Kuo et al., 2017; Mamoshina et al., 2018; Talukder et al., 2018).

In a Blockchain network, each user is considered as a Node and the transactions which occur are combined into Blocks. All blocks are connected with each other in a Chain. Each node contains at least one pair of Public-Private Keys. The public key is utilized to address the node as a sender or a receiver, whereas the private key is utilized by a sender (to sign the transaction being sent) and a receiver (to redeem the sent transaction upon receipt) (Nakamoto, 2008; Romano & Schmid, 2017; Zheng et al., 2017). In order to validate the key to decrypt and access the received data, and to conduct any further transactions, an agreement among the contributing nodes needs to be made in advance (Nakamoto, 2008; Romano & Schmid, 2017; Zheng et al., 2017). This ensures synchronization of all the copies of a Blockchain ledger throughout the entire network. On the occasion of any

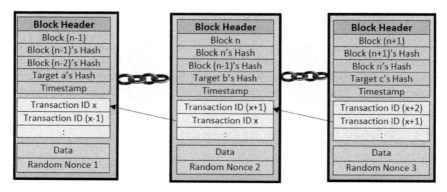

Figure 2.4. A conceptual representation of Blockchain (Shuaib et al., 2019, p. 3)

change in the chain, network nodes are notified. As all the blocks are chained, they are *private, anonymous, immutable, time-stamped* ledgers. Figure 2.4 illustrates a basic concept of Blockchain, where blocks are connected in a chain and the type of data a block may comprise.

The Block Header comprises data like a block number, hash value, the hash of the previous block, the targeted node's hash and a timestamp. Besides the header, a transactions list is also kept. This is an important element of Blockchain, as it comprises track records of any and all requests and valid transactions which have occurred in the network. In the end, some data may get included in the block, along with a randomized value (usually added to conduct "mining"). In mining, nodes are called "miners", which utilize their computational power to resolve numerical puzzles. The outcome values are used to conduct a consensus for executing a valid block (Romano & Schmid, 2017). The randomized value sought by the miners is utilized to execute a valid hash of the block, as proof that a certain measure or threshold is met before its inclusion into the hashing algorithm (Shuaib et al., 2019).

Blockchain's nature of decentralized distribution makes the network robust against any type of attack. As every node on the network comprises a copy of the ledger, data and transactions are *immutable* and *irrecoverable*. If any node of the network gets attacked or compromised, the network remains intact as the information in the attacked or compromised node exists in all other nodes of the network (Gupta, 2017). This restricts the illicit modification of data.

In general, there are three types of Blockchain: Private, Public, and Consortium (Hölbl et al., 2018; Zheng et al., 2017). Each type contains its own set of functionalities. Therefore, the selection of the type varies from case to case.

- **Private Blockchain:** This type of Blockchain network is suitable for organizations where employees are considered as nodes of the network and execute transactions. The network is restricted to organizational employees only. This type of Blockchain network can be denoted centralized, as the use is limited to the organization's employees (Hölbl et al., 2018; Zheng et al., 2017).

- **Public Blockchain:** As it is named, this type of network is available to the public. This type of network is *permission-less* (anyone can join the network and execute transactions) (Hölbl et al., 2018; Zheng et al., 2017). "Bitcoin" is a popular example of a public Blockchain network, where anyone can join the network and take part in Blockchain management (Shuaib et al., 2019).
- **Consortium Blockchain:** This is considered a permissioned public Blockchain due to its characteristic of limited availability to the public. This type of Blockchain is more open, compared to private Blockchain.In this type of network, a group of selected people (representing entities) takes part in transaction validation and/or Blockchain management. Based on permission rights (configured and executed by the administration), public users may be allowed to read permissions, but may not be privileged to take part in the consensus scheme applied to decision making (Zheng et al., 2017).

To settle an agreement in a Blockchain network, a consensus algorithm is applied over the valid block of transactions (Shuaib et al., 2019). Some commonly used consensus algorithms are:

- **Proof of Work (PoW):** This algorithm adopts the power of a node's Central Processing Unit (CPU) to participate and compete with other nodes in resolving a hash puzzle to regain a predetermined value (Gupta, 2017; Nakamoto, 2008). Successful execution of this rewards the node in the form of consensus power. This is determined by an assured amount of newly achieved cryptocurrency. The participating nodes or miners can perform this either alone or by forming a group. However, this algorithm is computationally expensive due to its nature of high energy consumption (Shuaib et al., 2019).
- **Proof-of-Stake (PoS):** This type of algorithm is based on the assets contained by a node, unlike the PoW (Gupta, 2017; Zheng et al., 2017). In a PoS based network, nodes with higher assets are likely to create and add new blocks in the chain, which may create discrimination in the network with the possibility of attacks from the poorer nodes (Shuaib et al., 2019). However, this algorithm consumes less computational power than PoW (Shuaib et al., 2019).
- **Proof of X (PoX):** This type of algorithm usually depends on qualification to determine which node comprises the ability to execute a new block to add in the chain (Sun, 2018). This type of algorithm is generally used in Public Blockchains, where nodes are needed to prove that they comprise a defined type of resource to take part in Blockchain activities. However, PoX-based Blockchain networks are risky (Shuaib et al., 2019). This is due to malicious nodes' presence, executing fake nodes in the network for claiming consensus power. In PoX-based Blockchain networks, the longest chains are considered trusted. But this does not ensure the prevention of unfair power as some nodes are lacking in resources to participate in decision making.

Consensus algorithms like Practical Byzantine Fault Tolerant (PFBT) can evade this by devising a pseudo-randomly designated leader, which determines a decision in accordance with each node's decisions (Sun, 2018; Wahab & Memood, 2018).

- **Practical Byzantine Fault Tolerance (PBFT):** This type of algorithm gets involved in settling disputes among the nodes of a Blockchain network. The aim of PBFT is to resolve Byzantine Generals Problem in a network, where applicable (Shuaib et al., 2019). This assumes the possibility that corrupt nodes exist in a network. Those corrupt nodes are responsible for executing wrong messages. It is assumed that 1/3 of the nodes in a network are corrupt (Blagojevic, 2019; Zheng et al., 2017). Therefore, to continue a transaction and add a new block in the chain, a node needs to achieve a 2/3 vote by the other nodes of the network (Blagojevic, 2019; Zheng et al., 2017).

Bitcoin cryptocurrency was implemented just after Nakamoto's white paper release over the Internet (Agbo et al., 2019). As the code was released open-source, numerous entities took the opportunity of improving this novel technology, resulting in different generations of Blockchain technologies. Bitcoin represents the very first generation of Blockchain (also known as Blockchain 1.0) (Swan, 2015). Some others are Dash[4], Monero[5], and Litecoin[6]. The second generation of Blockchain (Blockchain 2.0) is associated with smart contracts (programs which encode rules for smart properties) and smart properties (digital assets or properties whose ownership is manageable and controllable by a blockchain platform) (Swan, 2015). Examples of Blockchain 2.0 include NEO[7], Ethereum[8], Ethereum Classic[9], and QTUM[10]. The third generation of Blockchain (Blockchain 3.0) is focused on non-financial applications (Swan, 2015). Different industries worldwide are currently attempting to harness this technology's potential. Therefore, Blockchain is now admired as a general-purpose technology (Jovanovic & Rousseau, 2005).

Blockchain is also getting integrated into the digital healthcare ecosystem like other industries mentioned. Figure 2.5 provides a typical overview of Blockchain-based healthcare data management (Khezr et al., 2019):

Step 1: Primary data is executed based on patient and physician interaction. Data includes the current problem, health history and other related physiological information.

Step 2: Based on primary data, an EHR is created. The EHR also includes other patient data: medical images, drug history, and nursing care.

---

[4] Dash - Dash Is Digital Cash You Can Spend Anywhere (n.d.)
[5] Monero (n.d.)
[6] Litecoin - Open Source P2P Digital Currency (n.d.)
[7] NEO-project (n.d.)
[8] Home | Ethereum (n.d.)
[9] Ethereum Classic (n.d.)
[10] Home - Qtum (n.d.)

An Overview of Emerging Digital Health Informatics

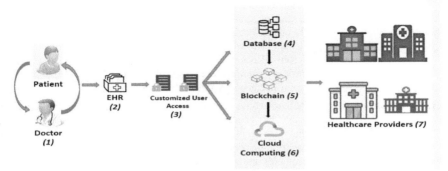

Figure 2.5. An overview of Blockchain-based healthcare data management (Khezr et al., 2019, p. 5)

Step 3: Customised Access Control is provided to the patient, who is the owner of sensitive information in the EHR. If any third party wants access to sensitive data, they must seek consent from the data owner. The owner is privileged to decide who is allowed to access his/her data and who is not.

Steps 4, 5 and 6: These steps perform the system's core functionality, which includes the traditional database, Blockchain and cloud computing. Database and cloud store the health records in a distributed manner, where Blockchain ensures Customised Access Control with high levels of privacy.

Step 7: Healthcare providers get access to the patient's data after consent from the patient for on-time healthcare delivery. Patient records are kept readily available to access (maintaining privacy and security) ensuring ubiquitous healthcare (Panesar, 2019).

The potential benefits of Blockchain for digital healthcare are summarized in Table 2.1 (Agbo et al., 2019).

Many research initiatives have been documented to ensure the mentioned benefits in today's digital Healthcare.

- Zyskind et al. (2015) proposed a Blockchain-based secure and privacy-preserving data sharing architecture for both service providers and mobile users. The proposed design executes two types of transactions (Tdata for ensuring data storage and retrieval, and Taccess for controlled access).
- Azaria et al. (2016) developed "MedRec", a Blockchain-based decentralized EMR management system which delivers a functional prototype for implementation. MedRec is established on the Ethereum platform and generates three types of Smart Contracts to assist in the storage of patients' medical information (stored with numerous healthcare providers), allowing third-party access upon authentic verification. The "Register" Contracts specifically maps a node's identity strings to its Ethereum address. The "Patient-Provider Relationship" (PPR) Contracts state the ownership

Table 2.1. Benefits of Blockchain for Digital Healthcare (Agbo et al., 2019)

Decentralization	By nature, healthcare is surrounded by distributed stakeholders who need a decentralized management system. Blockchain can fulfil the necessity by providing the desired system, where all stakeholders are privileged with controlled access without any central authority.
Health Data Ownership	Patients would like to: own their data, be aware of their data usage, have control over their data usage, and be assured that their data is in good hands. Blockchain is capable of ensuring these requirements through complex cryptographic protocols and well-structured smart contracts.
Improved Data Security and Privacy	All healthcare stakeholders are highly concerned and comply with the importance and compliance of data privacy, security and protection. Blockchain's "immutable" characteristics ensure high end data security (incorruptible, unalterable or irretrievable, time-stamped, affixed in chronological order and encrypted) for healthcare. Moreover, data stored in Blockchain uses cryptographic keys, which ensure privacy.
Transparency and Trust	Blockchain's nature of transparency creates a trustworthy environment for a distributed healthcare system, which enhances its applicability to healthcare stakeholders.
Robustness and Availability	Due to its distributed nature, a copy of patients' data is guaranteed to be available to multiple sources. Blockchain is resilient and robust against data corruption, data losses and several security attacks on data availability.
Auditability	Without accessing the patients' full record stored in Blockchain, integrity and validity of these records can be verified. This characteristic is very effective for healthcare, where auditability is an important requirement.

and stewardship of patients' medical data, including data to query strings (including data positions) and access permissions. The "Summary" Contracts comprise a list of PPR references, which define attachments with other patient nodes and those of healthcare provider. The system employs four software agents (Ethereum client, EMR manager, Backend library, and Database gatekeeper) on a system node to deploy the business logic of data sharing and management.

- Yue et al. (2016) proposed a Blockchain-based healthcare data gateway comprising access control policy (person-centric) to ensure data ownership, access control, and data sharing for patients, avoiding privacy violation. However, the mechanism lacks in providing an explanation on how the service conducts prevention without knowing the data content, as the computation is based on raw medical data.

- Based on MedRec, (Yang & Yang, 2018) approached a secure healthcare data sharing mechanism applying signcryption and attribute-based authentication. The proposed mechanism encrypts EHR data using a symmetric key, which is further encrypted by applying an attribute key set. The chain of both cipher-texts (encrypted EHR data and key) is signed using a private key. To perform data access, the user verifies a signature to execute decryption of EHR data using the key to obtain EHR plaintext.
- Zhao et al. (2017) utilized fuzzy vault technology for designing a recovery scheme and a lightweight backup to manage keys (used for encrypting health signal data obtained from a Body Sensor Network (BSN)) and storing it in a health Blockchain. However, the work lacks in defining how health Blockchain works.
- Xia et al. (2017) developed BBDS, a high-level Blockchain-based framework that allows data owners and users to access health records from a shared data repository upon verification of identities and keys. They applied an identity-based authentication and key agreement protocol to obtain the authentication of users' membership. Nevertheless, they have limited access to sharing of sensitive health information to verified and invited users only.
- Xia et al. (2017) also developed MedShare, a Blockchain-based framework for healthcare data sharing. The developed framework provides data auditing, provenance and control in cloud-based data repositories among healthcare providers.
- Liang et al. (2017) developed a Blockchain-based (permissioned) user-centric framework for personal health data sharing. The framework has adopted a channel formation scheme and Hyperledger Fabric membership service to provide identity management and privacy protection. Moreover, they deployed a smartphone app for health data collection from wearable devices. The collected data is synchronized in cloud storage, allowing data sharing with healthcare providers.
- Ahram et al. (2017) developed HealthChain, a permissioned-private Blockchain-based EMR application. The application is employed on Bluemix and uses Hyperledger Fabric Blockchain. The modular architecture of IBM's Hyperledger Fabric (Androulaki et al., 2018) ensures data security, confidentiality, and scalability for HealthChain. The application also deploys Chain codes (Smart Contracts) to assure controlled access and authorization privilege in the network.
- Omar et al. (2017) introduced MediBchain, a privacy-preserving platform for Blockchain-based EMR systems using encryption functions to de-identify patients' data.
- Kuo and Ohno-Machado (2018) developed "Modelchain", a Blockchain-based mechanism for privacy-preserving machine learning. The mechanism is purposed to expedite healthcare research and assist in quality enhancements.

In its design, they employed a proof-of-information algorithm (above the PoW consensus protocol) to execute machine learning (online) to enhance accuracy and efficiency.

- Fan et al. (2018) developed MedBlock, a Blockchain-based hybrid architecture to secure EMRs. In the architecture, the nodes are divided into three types: orderers, endorsers and committers. The consensus protocol used is a different type of PBFT (Castro & Liskov, 2002) consensus protocol. However, the proposed architecture lacks an explanation of the imposed access control policy, which enables third-party researchers to access data. Furthermore, the applied asymmetric encryption algorithm is not wise to adopt because of performance issues (encryption/decryption) (Jin et al., 2019).
- Wang et al. (2018) developed a parallel healthcare system (PHS), where the achievement of perspective, descriptive and predictive intelligence is based on parallel execution, computational experiments and artificial systems. A consortium blockchain is implemented (contains information on patients, healthcare providers, health bureau and communities, and healthcare researchers). The framework deploys smart contracts to review, share, and audit healthcare records.
- Zhang and Lin (2018) developed "BSPP", a hybrid Blockchain-Based Secure and Privacy-Preserving patient health information (PHI) sharing scheme. The scheme uses a private Blockchain (to store PHI for individual hospitals) and a Consortium Blockchain (to securely retain the PHI indices). The scheme has adopted a public key encryption (with keyword search) scheme (Boneh et al., 2004) to secure the PHI search and assure identity privacy.
- Guo, et al. (2018) developed an attribute-based signature scheme for blockchain encapsulated EHR. The scheme allows message endorsement by the patient in accordance with the attribute, without disclosing any information (except the endorsed message). The scheme allows multiple authorities' access and performs efficiently (proved in comparison studies).
- Li et al. (2018) developed a Data Preservation System (DPS) for EMR based on the Ethereum Blockchain platform.
- Cichosz et al. (2019) proposed architecture using multi-signature Blockchain contracts for health data sharing of diabetes patients to ensure data privacy and access control.
- Patel (2019) developed a Blockchain-based decentralised framework which enables image sharing among different healthcare entities with customized user permission. The framework assists in the overall development of radiological studies.
- Shen et al. (2019) developed MedChain, an efficient session-based architecture for health data sharing using Blockchain. It employs a digest chain structure approach to verify a shared health IoT data stream's reliability. MedChain was developed to overcome the efficiency issues of existing solutions like Medblock (Fan et al., 2018) and MedRec (Azaria et al., 2016).

The experimental results show that MedChain satisfactorily ensures higher efficiency in fulfilling data sharing security requirements.

Besides the mentioned research, several Blockchain implementation initiatives from government and non-governmental organizations are also found in the literature. Guardtime (an Estonian company) applied a Blockchain-based platform to store and secure over one million Estonian patients' health records (Angraal et al., 2017; Mettler, 2016). Healthbank (a digital health company from Switzerland) adopted a blockchain platform for ensuring controlled access by patients to their data (Azaria et al., 2016). Medicalchain, a Blockchain-based platform was developed to facilitate cross-border patients' medical record sharing. The platform is also aimed towards a Healthcoin initiative (influenced by Bitcoin) and global EMR system (Engelhardt, 2017). HealthNautica (a US company) collaborates with Factom (a US-based Blockchain company), utilizing DLT for securely storing medical information to enhance audit trail efficiency (Suveen et al., 2017).

In addition to research on data sharing, storage, security, privacy-preservation, and encryption, several research findings in the literature address patient-centric remote monitoring using Blockchain. Uddin et al. (2018) proposed a Patient-Centric Agent (PCA) based on Blockchain to ensure data security and privacy (end to end) in a remote patient monitoring application. Firdaus et al. (2018) introduced a Blockchain-based practical swarm optimization (PSO) method for feature optimization and root exploitation detection in a mobile device based health data management system/scheme. Ji et al. (2018) developed a Blockchain-based Multilevel Privacy-Preserving Location Sharing (BMPLS) scheme for remote monitoring applications. The goal of the scheme is to ensure privacy-preserving location sharing for telemedical information systems using Blockchain. The scheme defines the primary requirements for location sharing confidentiality, variability, decentralization, multilevel privacy protection, irreversibility, and tampering evidence by applying order-preserving cryptography and Merkle trees. The scheme's evolution proves its practicality and feasibility for users (patients and physicians). The scheme can be integrated into telemedical information systems for location information protection. Hardjono and Pentland (2019) developed "ChainAnchor", a Blockchain-based (permissioned) identity verification system which offers an anonymous identity verification service to transaction performing nodes of the network. The system adopts an Enhanced Privacy ID (EPID) scheme (a zero-knowledge proof scheme) to assure the nodes' anonymity in the network.

Several research findings in the literature address Blockchain in a Cloud environment. Chen et al. (2018) proposed a Blockchain framework with secure cloud storage for healthcare data sharing to secure patients' sensitive medical records. In the framework, healthcare data management is realized through a digital archive, which contains access control rights and information on its owners. The data is stored in the chain by employing cloud encryption. Wang and Song (2018) developed a cloud-based secure EHR system using an attribute-based cryptosystem and Blockchain. The system uses a blend of identity-based cryptography and signature to encrypt medical data and implement digital signs. To assure the traceability and integrity of healthcare facilities, the system employed different techniques on top of Blockchain. Kaur et al. (2018) developed BlockCloud, a cloud-based healthcare data management system with Blockchain. The reason for deploying both technologies together is to ensure secure data distribution without any third party involvement. The study addressed the challenges of collaboration among the healthcare entities (public health agencies, healthcare providers, governments) creating and enforcing healthcare policies (Kaur et al., 2018; Li et al., 2019).

Zhu et al. (2019) developed a controllable Blockchain-based data management approach in a cloud context. The model addresses user concerns regarding the lack of control found in the posted ledger. It adopts a special trust authority node which enables the user to prevent and terminate any impending malicious action. Omar et al. (2019) designed a Cloud-based patient-centric health data management system using Blockchain (as storage) to achieve privacy. The system executes a set of security and privacy requirements to ensure integrity, accountability, and security for healthcare data stored on a Blockchain.

Interoperability is one of the most prominent challenges in today's digital healthcare ecosystem. This issue is addressed in a different Blockchain focused research.

- Peterson et al. (2016) proposed a cross-institutional healthcare information sharing approach based on Blockchain. The design introduces a new transaction and block structure to ensure secure access to FHIR (Fast Healthcare Interoperability Resources) transactions stored off-chain. A new consensus algorithm was introduced, eliminating the necessity of expensive resources (e.g. PoW consensus algorithm). In the design, a block needs to go through several phases (transaction distribution phase, block verification request phase, signed block return phase, and Blockchain distribution phase) before getting added to the Blockchain. The design also proposed a proof-of-interoperability concept for the transacted data to ensure compliance with FHIR's semantic and structural limitations. A random miner selection algorithm is used to ensure equal opportunity for the nodes to become a miner in future events. However, the proposed approach does not provide any information about the organization, storage, and medical data access. In addition, the adopted privacy-preserving keyword search mechanism lacks algorithm detail.
- McFarlane et al. (2017) proposed "Patientory", a Blockchain-based peer to peer EMR network for storage (HIPAA compliant). The researchers also developed a software framework to address access control, authorization, authentication, and data encryption during system implementation. The software framework is also used for token management and interoperability improvement.

An Overview of Emerging Digital Health Informatics **35**

- Roehrs et al. (2017) developed OmniPHR, a distributed model to integrate PHR in a parallel database for keeping it in blocks. The model also combines structural-semantic interoperability and numerous updated PHR formats.
- Dagher et al. (2018) developed Ancile, an Ethereum based Blockchain platform to ensure data security, privacy, access control, and interoperability of EMR.
- Jiang et al. (2018) proposed BlockHIE, a Blockchain-based Health Information Exchange platform to ensure secure health information exchange and storage in healthcare entities.
- Zhang et al. (2018) proposed FHIRChain, a Blockchain-based architecture compressing HL7 (Health Level Seven) FHIR standard for secure health information exchange among healthcare entities.

Blockchain (along with other emerging technologies) offers a unique opportunity to harness its potential for the advancement of personalized healthcare services (Shae & Tsai, 2018). Several studies are published addressing this opportunity.

- BNA.Bits (an Israeli company) plans to use Blockchain for storing medical and genetic data in a personal "DNA wallet". The concept enables access to medical and genetic data by relevant researchers with data privacy and security (Kuo et al., 2017).
- Mamoshina et al. (2018) introduced a comprehensive roadmap describing the implication of emerging technologies in precision healthcare.
- Hussein et al. (2018) proposed a Blockchain-based framework to secure medical records by using discrete wavelet transforms and genetic algorithms. The framework deploys a customized cryptographic hash generator to execute the required user security key. Moreover, it uses MD5 (a hash function based message-digest algorithm which is able to generate 128-bit hash value strings ("RFA-MD-15-013: NIMHD Transdisciplinary Collaborative Centres for Health Disparities Research Focused on Precision Medicine (U54)," 1992)) to execute a new key format by implementing a discrete wavelet transform. This framework improves the general system's security and robustness against malicious attacks.
- Juneja and Marefat (2018) conducted an experiment using Blockchain and deep learning architecture to determine arrhythmia classification.
- BlockVerify (a UK based company) employs Blockchain to fight against imitation drugs by securely attaching a unique verification code to drug packages. The attached tags can be removed after verification of drug legitimacy through Blockchain (Roman-Belmonte et al., 2018).
- Ethereum (www.ethereum.org) introduced Vibranthealthchain, a Blockchain platform deployed in the cloud. The platform uses Medical Miner (MM) to interpret clinical notes in accordance with structured International Classification of Disease (ICD) and SNOMED CT codes, to validate episodic transactions. The ICD and SNOMED CT codes are preserved in

smart contracts. In terms of genomic data, MM employs the proprietary tools (Talukder et al., 2018) to execute genomic mutation and transform it into the Human Genome Variation Society (HGVS) taxonomy. During this execution, the healthcare specialist authenticates whether the existing onset matches any other clinical pathways (Kinsman et al., 2010).

2.4 Data Analytics and Modelling

This category includes technologies such as Virtual Reality (VR), Augmented Reality (AR), Artificial Intelligence (AI), Machine Learning (ML), Deep Learning (DL), Digital Twin, and Big Data; which are usually used for data analytics and modelling. The description covers definition and prior research on implications for patient-centric healthcare for all mentioned technologies. The working mechanism and characteristics of each class of emerging technologies are beyond this description's scope due to variations in algorithm, method, and execution procedures.

2.4.1 Artificial Intelligence (AI) and Machine Learning (ML)

AI is an elite branch of computer science (Nils J. Nilsson, 1998). It aims to develop new machine intelligence which can respond and communicate like human intelligence (Lawrence et al., 2016). ML can be defined as a subset of AI, which strategically applies algorithms to synthesize the cardinality between data and information (Awad & Khanna, 2015). Unlike AI, ML comprises an adaptive approach (for example, neural networks) to get training on new data without any extensive coding of new rules (standard prerequisite of AI algorithms such as expert systems) (Nilsson, 1980; Russell & Norvig, 2009).

Research in AI and ML mostly focused on language and image recognition, natural language processing (NLP), expert systems and robotics (Liu & Tang, 2018). Since the inception of these technologies, it has advanced at a rapid pace and is gradually being employed in the healthcare industry (Grey, 2016; Handelman et al., 2018; Kantarjian & Yu, 2015; Krittanawong et al., 2018; Lawrence et al., 2016). Recent studies of these emerging technologies in digital health are mostly focusing on personalized healthcare services.

- IBM launched "Watson for Oncology", a special program to provide enhanced evidence-based treatment options to clinicians (Zauderer et al., 2014). The program comprises advanced analysis ability (empowered by AI) to interpret the meaning and context of data (structured or unstructured) in clinical reports and notes, which is otherwise difficult to interpret for selecting the right treatment pathway. Furthermore, it combines the interpreted data with the patient's traditional file data to determine a patient's treatment plan.
- Guidi et al. (2015) designed an AI-based android app for bipolar disorder patients. The app performs the analysis of prosodic features in running speech. It records audio samples and evaluates speech frequency (including

An Overview of Emerging Digital Health Informatics **37**

changes). The app transfers the accumulated patient data to a central server for further processing. The app's audio recording quality and algorithm was evaluated (consistency and performance) in an experimental study and found effective.

- Tseng et al. (2016) developed an AI-based interactive system to provide assistance to patients with chronic diseases (e.g. metabolic syndrome). The system evaluates routine check-up results, assesses potential risks and delivers personalized healthcare services (exercise and diet guidelines) to the patients. The system implies interactive procedures to accumulate patient's feedback on vital signs and provides real-time suggestions on health management. The system is fully operational with a computer or smartphone.
- IBM launched another program called "Medical Sieve" to design and develop a "cognitive assistant" for next-generation healthcare with reasoning analytics, reasoning capabilities, and a data bank of clinical knowledge (Syeda-Mahmood et al., 2016). The program is proficient in supporting clinical decision-making for cardiology and radiology. The "cognitive assistant" is capable of analyzing radiological images with accuracy and in an expedient manner.
- Alaa et al. (2016) introduced "ConfidentCare", a Clinical Decision Support System (CDSS) empowered by AI. The system studies personalized screening policies from EHR and learns the "best" screening procedure using a supervised learning (machine learning) algorithm. The algorithm authenticates the learned screening policy with predefined accuracy requirements. The experimental result shows that it outperforms the existing CPGs (Clinical Practice Guidelines) in false-positive rates and cost.
- Deep Genomics (www.deepgenomics.com) is purposed to assist in mutation detection and disease linkage discovery by identifying patterns in medical data banks and genomic information. Their research is currently focusing on emerging disruptive technologies capable of informing physicians about the cause and effect of DNA alteration (Mesko, 2017).
- Daowd et al. (2018) introduced "PRISM", a digital health platform based on AI, data visualization tools and mobile health technologies. The platform aims to empower patients with self-monitoring, self-assessment and self-management of chronic conditions. PRISM provides risk assessment tools for chronic conditions (five types), psychiatric disorders (two types), and cancers (eight types).
- De Silva et al. (2018) developed an AI-based analytics framework by employing natural language processing and machine learning techniques to execute automated aggregation of patients' information and intelligent analysis. The framework is capable of identifying conversation routes in different online support groups. Moreover, it is able to detect, extract, and analyze patient interaction, behavior, emotions, demographics, clinical factors, and decisions. The framework is purposed to inspect the impact

- of online social influences on patients regarding treatment selection, rehabilitation, side effects and the emotions expressed on a regular basis.
- Google Brain and Verily Life Science developed "DeepVariant", an AI-empowered visualization tool. The tool is able to construct an accurate image of a person's genome based on his/her sequencing data (Qu et al., 2019). The tool complements data with a deep learning system, fed by millions of detailed reads and complete genome sequences. The tool comprehensively tweaks its own parameters until it starts to understand the sequenced data with efficiency. The tool won the PrecisionFDA Truth Challenge award in 2016 for the highest SNP (single-nucleotide polymorphism) performance (Qu et al., 2019).
- Baek et al. (2019) introduced a context mining based mental health model, empowered by AI. The model is based on the users' profiles from depression and the health weather index of the Korean Meteorological Administration. This has been done to categorize and define semantic ontology based on context information and to construct a context mining model for the depression index service. The model utilizes personalized context information to deliver the indexed depression service. The model also utilizes user-based information to perform modelling to deliver guidelines for constructing a depression data model. The model is found to be efficient in delivering precise services to users along with customization.
- Chen and Zheng (2015) developed a machine learning and similarity calculation algorithm-based application scheme for primary and home healthcare. The scheme follows the patient's movement and obtains physiological data. The application scheme also provides medical services such as doctor's recommendations.
- Hijazi et al. (2016) proposed a machine learning-based method to filter patients' ECGs (electrocardiograms). The proposed method applies a machine learning classifier for cardiac health risk detection and evaluating severity.
- Castro et al. (2017) proposed a Human Activity Recognition (HAR) system based on IoT to monitor a user's vital signs remotely. The system applies ML algorithms to execute the accomplished activity in four pre-set classes (sit, walk, jog, and lie). Using remote monitoring components, the system delivers feedback on activities. The system also offers remote visualization and alarms (customizable). In the experimental study, the system resulted in a ratio of 95.83% accuracy.
- Alfian et al. (2018) proposed a personalized healthcare system based on ML. The system provides assistance to diabetic patients in self-management. The system employs a Bluetooth Low Energy (BLE) sensor to gather vital sign data such as blood glucose, blood pressure, heart rate, and weight. The collected sensory data is transferred to a smartphone for real-time data processing. The system employs MongoDB (a cross-platform document-oriented database program) for data storage and Apache Kafka (an open-source stream-processing software) to facilitate the real-time data processing.

The experimental results show that the proposed system is adequately efficient in performance compared to commercially available solutions.

- Öksüz et al. (2018) proposed a Data-analytical Information System called "DAIS". The system is able to forecast future Body Mass Index (BMI) changes before initiating any therapy. It considers the existing parameters (heart rate, age) through standard exercises. The system employs ML models to analyze physical data (accompanied by a user's randomized clinical trial (RCT) data) and delivering prediction on change. On demonstration, the system performed efficiently with an accuracy of 85%.
- Magyar et al. (2019) developed a prototype called "VASelfCare" to assist elderly patients (with type 2 diabetes mellitus) in self-care. The prototype adopts machine learning and artificial intelligence techniques to provide virtual assistance to the patients in line with their preferences and emotions.
- Bali et al. (2019) developed "Diabot" (DIAgnostic chatBOT), a generic text to text chatbot which conducts conversations with patients by using NLU (Natural Language Understanding) techniques. The chatbot aims to deliver personalized predictions by utilizing the patient's general health data and vital signs. The chatbot has since been extended to "DIAbetes chatBOT" providing specialized diabetes predictions using the Pima Indian diabetes dataset. To determine predictions, it uses multiple ML-based classification algorithms (applied based on accuracy).
- Amin et al. (2019) proposed a real-time analytics approach to monitor the vital signs of a patient. The approach fully depends on seamless vital sign data derived from sensors. It provides notification to the patient if any risky symptom of cardiovascular disease is observed. Furthermore, researchers developed an ML model to execute predictions based on the collected physiological data. This approach's deliverable is a personalized healthcare service, with improved quality care.

2.4.2 Big Data (BD)

BD is a term that deals with the massive volume of complex datasets, where traditional data processing methods are ineffectual (He et al., 2017). To date, there is no specific definition of BD (Alonso et al., 2017; Ketchersid, 2013; Olivera et al., 2019). BD was first introduced in 1997. It was used for data analytics for the first time in 2001 and it emerged with a data explosion phenomenon (growing volume, variety and velocity of data) in different sectors including healthcare (Laney, 2001). In literature, various disparate BD descriptions are found (Baro et al., 2015), but most of them agree that it comprises 5Vs: Volume, Velocity, Variety, Veracity, and Value (Huang et al., 2016). Therefore, it is understood that BD definition is subject to technological advancement (He et al., 2017).

In BD empowered digital health (He et al., 2017), "Volume" represents the massive quantity of patient records contributed by EHRs. IIt comprises medical records such as medical and treatment history, patient information, neuroimaging data, and various large-scale medical images. "Velocity" represents the data collection at high speed through patient monitoring in real-time. "Variety" represents the numerous types of datasets from different sources. "Veracity" represents the data quality and evidence extracted from the collected data. "Value" represents the usefulness of data in improving healthcare outcomes.

BD is playing a significant role in the transformation of conventional healthcare to digital health. Due to the competences and potential, BD is being widely accepted nowadays in the healthcare entities (Khan et al., 2014; Klonoff, 2013). Recent studies on BD are mostly found focusing on personalized healthcare. For instance:

- Fang et al. (2016) proposed a framework for big healthcare data. The proposed systematic data-processing pipeline framework provides features such as data searching, capturing, analyzing, storing, decision support, and sharing. The framework accumulates genomic, behavior, EHRs, and public health data. For data analyses, the framework applies machine learning algorithms and feature selection.
- Sakr and Elgammal (2016) proposed a smart health framework. It offers data analytics for smart healthcare applications. The framework accumulates patient data from different resources (e.g. hospital information systems, laboratory and radiology information systems). It employs predictive modelling and pattern matching techniques to offer data analytics service. The framework incorporates cloud computing, big data and integrated sensor technologies to provide enhanced healthcare services.
- Mahmud et al. (2016) proposed an analytics and data visualization framework for health shock-prediction. The system accumulates data from different geographic, socio-economic and cultural contexts. It employs AWS[11] (Amazon Web Services) and a geographical information system (GIS) for data collection, indexing, storage, and visualization. The framework also employs predictive modelling (including summarized fuzzy rules) for data analysis.
- Hossain and Muhammad (2016) introduced a cloud-based Big Data framework to provide voice pathology assessment (VPA). The system applies MPEG-7 low-level audio and entangled derivative pattern for speech or voice signal processing. It employs a Gaussian mixture model (as a classifier) and ML algorithms (as support) to a vector machine[12]. The system demonstrated well in accuracy and time requirement in the experimental study.
- Zhang et al. (2017) proposed "Health-CPS", a cyber-physical system based on cloud and Big Data analytics. The system aims to provide personalized healthcare services and applications. It incorporates a data collection layer (to retain the unified standard), a data-oriented service layer, and a data

---

[11] A cloud computing service.
[12] An extreme learning machine.

management layer (to support parallel computing and distributed storage). In the prototype demonstration, the system performed satisfactorily and can be employed for different personalized smart healthcare services and applications.

- Hossain et al. (2017) proposed "T-CPS", a cloud-oriented and Big Data assisted energy-aware cyber-physical therapy system. The system incorporates smart devices and smart things for therapy sense. The framework uses multimodal sensing for therapy sensing, visualization, annotation, therapy playback, and energy efficacy.
- Pramanik et al. (2017) introduced a big data-enabled smart healthcare system framework (BSHSF). The framework provides a service-oriented infrastructure to deliver smart healthcare services. It accumulates diagnosis reporting, EHR, biometric, social media and surveillance data. The framework also facilitates healthcare knowledge discovery by incorporating a smart service infrastructure mechanism.
- Shaban-Nejad et al. (2017) introduced a semantic web application called "PopHR" (Population Health Record). The application automates integration and extraction of enormous heterogeneous data (Big Data), contributed by numerous distributed sources (clinical records, administrative data, survey responses). The platform aims to support population health monitoring, evaluate system performance, and enhance decision making in the design, deployment, and assessment of system interventions and population health.
- Sahoo et al. (2018) introduced big data analytics and a computing model based on service-level agreement (SLA). The model applied a PSNB (Parallel Semi-Naive Bayes) based probabilistic method to process Big healthcare data stored in the cloud to predict health conditions. The model uses an MCA (Modified Conjunctive Attribute) algorithm to improve the efficiency of PSNB method. The demonstration method performed efficiently with higher accuracy (87.8%).
- Sinnapolu and Alawneh (2018) developed a big data analytics-based health monitoring and emergency assistance model. The proposed model works with two iOS apps (healthdetect and healthlocateapp) simultaneously. The healthdetect app is used to monitor heart rate (uses a proximity sensor to detect heart rate) and the healthlocateapp is used to locate and route the nearest hospital (using a microcontroller in the vehicle) in emergencies.
- Kafhali and Salah (2018) proposed a Big Data analytics model based on a network of queues to illustrate performance modelling for healthcare monitoring systems. The model is able to estimate the minimum number of required computing resources to meet a SLA. In the experimental study, the proposed model can predict system response times and precisely determine the needed number of computing resources for healthcare data services to achieve the anticipated performance.
- Almogren (2018) developed a Big Data analytics-based automated and intelligent Parkinson disease (PD) monitoring system. The system uses

wearable devices and cloud technology to gather PD information (gait information, voice samples). The system enhances in-house monitoring capability by detecting early symptoms of PD. The system can assess voice disorders or dysphonia of PD patients, which enables physicians to recognize PD symptoms in patients. On demonstration, the system performed better in accuracy compared to other approaches.

- Hussein et al. (2018) proposed a remote cloud-based automated heart rate variability monitoring system. The proposed system uses big data analytics to determine patients' health status. It was developed to serve patients residing in remote areas. In the experimental study, the proposed system resulted in higher sensitivity (98.78%), accuracy (99.02%) and specificity (99.17%), signifying its robustness, reliability and value.
- Jain et al. (2019) proposed a Big Data-oriented readmission risk analysis framework. The framework is based on High-Performance Computing Cluster (HPCC). It includes a Naive Bayes classification algorithm. The framework shows that it can significantly decrease the overall assessment time with Big Data analytics and parallel computing platform (without affecting the performance).
- BD2Decide project (Lopez-Perez et al., 2019) presented a Big Data integrated DSS (Decision Support System). The system is able to link data retrieved from numerous disciplines. It provides important information for tailored treatment and care pathways for head and neck cancer patients.

2.4.3   Virtual Reality and Augmented Reality (VR and AR)

VR can be defined as an interactive simulation generated by a computer. It engages the user in such an environment that it feels like the real-world (Weiss et al., 2006). It is a progressive form of HCI (human-computer interaction) (Choi et al., 2015). The most recent form of VR is termed as augmented reality (AR) (Lee & Lee, 2018). In simple words, AR is a phenomenon where real-world objects are enhanced in an acknowledged context with complementary graphical information to assist the augmentation process (Adenuga et al., 2019). It delivers three basic features: the arrangement of virtual and real-world, real-time interaction, and accurate 3D (three dimensions) registration of virtual and real objects (Wu et al., 2013). The sensory information generated by an AR system can be constructive or destructive (either improving or masking the real-world context) (Rosenberg, 1992). An AR system can modify an ongoing perception of a real-world environment, whereas VR substitutes the real-world environment with a simulated version (Steuer, 1992). In healthcare, AR can provide precision and higher accuracy in disease diagnosis and surgical procedures with reduced medical errors and incidents (Adenuga et al., 2019). Several research findings on VR and AR in patient-centric healthcare have been found in recent studies.

- Paraskevopoulos and Tsekleves (2013) presented a motion capture system for Parkinson's Disease (PD) rehabilitation. The system employs commercial

An Overview of Emerging Digital Health Informatics **43**

game consoles and uses tailored video games to achieve PD rehabilitation protocol compliance. The system is able to detect random upper limb, unilateral or bilateral movements. It transfers the captured data to an online VR instructor platform and provides live performance feedback. The system enables clinicians to adjust game difficulty level remotely.

- ImmersiveTouch (an AR platform of the University of Illinois, Chicago, US) introduced different surgical training modules for percutaneous spinal screw placement, ventriculostomy, aneurysm clipping, and other various techniques (Alaraj et al., 2015; Luciano et al., 2013; Yudkowsky et al., 2013). The architecture provides a high-resolution stereoscopic view with a detailed and realistic surgical simulation.
- Spanakis et al. (2014) presented "MyHealthAvatar" (MHA), a digital bag for personal health data collection. It aims to provide meaningful and sustainable health information collected over a lifetime. MHA acts as an inimitable digital companion by seamlessly following and empowering patients through health-related services. MHA deals with various types of data such as genetic data, clinical data, activity data, behavior data, and sensory data; which can be utilized for the patient's health analysis, and prediction and prevention of diseases.
- Kohout and Kukačka (2014) proposed a VR based automated, real-time method for modelling an indiscriminately large number of muscle fibres. The method is able to represent the same muscle volume as on a surface mesh. It is based on an iterative transmuting of a predefined fibre template into muscle volume. The method is suitable for virtual physiological human (VPH) purpose and can be used as an interactive medical education software.
- Hsiao and Rashvand (2015) presented a mobile AR system for both physical and cognitive rehabilitation. The system provides embedded exercise guidelines to its users. The system rewards its users with virtual tickets, which are used to unlock and advance to a higher difficulty level.
- OrCam (www.orcam.com) developed a portable device for artificial visioning (Alaraj et al., 2015). The device is based on a mini television camera and can be mounted on the spectacles frame. It uses a microchip for optical character recognition (Moisseiev & Mannis, 2016). The device is able to detect texts and patterns (including face). The devices apply algorithm to convert the detected item(s) into word(s). The device also provides verbal translation to its users.
- Researchers at the University of Houston have developed a setup called "cave" to ease and support the drug addiction rehabilitation processes (Loria, 2016). The setup uses goggles to transform the wall projected images of a room (cave) into a Three Dimensional High Definition (3DHD) experience.
- Microsoft Inc. developed an AR headset called "HoloLens" (www.microsoft. com/en-us/hololens), which provides a realistic virtual experience in the demonstrating environment. It uses visor to illustrate high-quality 3D images

on the visual surface. The headset provides "calibration" to its users as an adjusting feature. It also uses 3D speakers to deliver binaural audio on the virtual surface (at any point). The key feature of HoloLens is that it does not interfere with the user's hearing and view of the real-world environment (Golab et al., 2016).

- Codreanu and Florea (2016) proposed a VR based home system that adapts the serious games approach for aged patients. The system uses machine learning models for remote activity monitoring and exercise detection. It aims to decrease physical effort. Through a motivating and realistic virtual world, the patient initiates kinesiology exercises, takes part in quizzes and delivers feedback. It accumulates the patient's exercise data, interprets, and delivers personalized care models. The system develops user profile, adjusts the game's difficulty level, evaluates the daily exercise progress, and provides performance feedback. The system is able to learn the user's behavior by detecting his/her symptoms and falls.
- Lv et al. (2017) designed a VR based training tool to assist dysphonia rehabilitation. The tool uses the built-in microphone of a tablet and/or Kinect2[13] for voice input. As an attractive logic part, the tool employs serious games (a game designed for a specific purpose other than entertainment). It allows users to play the game without any interventions and facilitates voice training in accordance with the therapist's guideline simultaneously. The tool also records the user's voice and provides extraction for long term rehabilitation progress evaluation.
- 3D Systems Inc. (www.3dsystems.com) provides various virtual endovascular training modules, especially for aneurysm and acute stroke. The modules are capable of delivering haptic feedback, which can be useful for decreasing the patient exposure to fluoroscopy during surgical procedures and improving the overall surgical efficiency (Kim et al., 2016; Saratzis et al., 2017; Velu et al., 2017).
- Google Inc. introduced Google Glass (www.google.com/glass), a combination of a projector, prism screen, and a minicomputer in a spectacles pair (Golab et al., 2016; Lee et al., 2017; Yoon et al., 2017). It detects objects thru neuro-monitoring or navigation and provides a visual on the surgeon's glasses directly. It also eradicates the necessity of looking away from the surgical context (Yoon et al., 2017).
- Phillips' engineering wing (Bothell, WA) introduced an AR-based solution by conjoining optical imaging with 3D X-ray images (Lee & Lee, 2018). The solution provides a distinctive AR view of a patient's spine, assists in planning the surgical path, and places pedicle screws consequently during the surgical procedures using AR navigation (automatic). This solution eradicates the need for a CT scan and decreases medicine doses.

---

[13] An input device for motion sensing from Microsoft.

- Debarba et al. (2018) proposed an AR-based visualization tool for human motion analysis. Their previous research on joint biomechanical modelling for kinematic analysis was the outcome of their previous research on personalized anatomical reconstruction of joint structures and optical motions extracted from medical imaging. The tool provides in situ visualizations of joint movements to the healthcare professionals.
- Similarly, Debarba et al. (2018) proposed a visualization tool of joint movements for sports medicine and rehabilitation based on augmented reality. The tool contains a recording mechanism to evaluate the acquired movements and different motion-related information. The tool uses the evaluation outcome to determine the patient's progress in kinematics thru the rehabilitation phase.
- The Ohio Centre of Excellence in Knowledge-enabled Computing (Kno.e.sis) introduced "kHealth–ADHF" (Active Decompensated Heart Failure) mobile app, an augmented reality-based personalized healthcare (APH) framework to provide health fitness and wellbeing with informed decision making to the users. The app asks targeted questions (developed with cardiovascular knowledge) to its users for seamless monitoring. It uses sensors to collect patient data: blood pressure, heart rate, and weight by analysing the user answers and sensor data. The centre also introduced another AR-based app called "kHealth–asthma" for asthmatic children. The app collects internet surfing data and social media data, and physical data to understand its health status better. Both the applications deal with enormous multimodal and multisensory data and face the Big Data challenge. To overcome this challenge, both the apps employ the SSN (Semantic Sensor Network) ontology to maintain contextual integration, annotation (knowledge graph) and interpretation of sensor data (Sheth et al., 2018).
- Palanica et al. (2019) developed "HealthVoyager", a VR technology to provide assistance in knowledge transfer between physician and patient. HealthVoyager consists of a VR software system (customizable), which can be used in smartphones or tablets. It can demonstrate personalized procedural and surgical findings along with normal anatomy images. The technology is capable of enhancing healthcare understanding of patients relevant to various disease states.
- Fonseca et al. (2019) developed "VR4Neuropain", a VR based system to monitor electrophysiological data in real-time. The system includes three key components: VR interface, a Glove (GNeuroPathy), and a Platform. It enables occupational therapists, physiotherapists, and physicians to determine interactive and innovative intervention procedures.
- Korn et al. (2019) developed an augmented reality-based system to provide enhanced rehabilitation to aged people. The system includes a micro projector (worn on the user's belt) to provide user instructions. The system also uses LEDs to illustrate training exercises (gamified) on the surface.

2.4.4 Digital Twin (DT): Integrating Multiple Technologies

Digital Twining is a recent innovation which integrates multiple digital health technologies such as IoT, big data, VR-AR, AI and data analytics. It is a simulation approach for a healthcare system or a healthcare activity (Liu et al., 2019). It applies multi-science, multi-physics and multi-scale models (Glaessgen & Stargel, 2012) to deliver expressed, accurate and efficient healthcare service (Liu et al., 2019). Generally, it comprises three components: (i) virtual object, (ii) physical object and (iii) health data (Liu et al., 2019). The virtual object contains a digital patient, medical device, wearables, external factor, and digital system model. The physical object contains a patient, a medical or wearable device, an external factor (social behavior, government policy, weather which may influence a patient's health) or a system comprising several or all of these objects. The health data contains monitoring data (transmitted by wearables in real-time), medical history and records (contributed by the healthcare providers), detection (contributed by the medical devices or external systems), simulation (presented by the digital models), and service data (contributed by the service platforms or systems).

DT comprises the potential of transforming the existing healthcare paradigm. Recent studies on this revolutionary technology are focussed on personalized healthcare services. For instance:

- Liu et al. (2019) introduced a cloud-based digital twin healthcare system called "CloudDTH". The system is able to diagnose, monitor, and predict various health aspects of a user. The system applies different wearables to accumulate user data. The system is an example of a cloud-based, generalized, extensible and novel framework. It aims to deliver customized health management services specifically to aged users.
- Martinez-Velazquez et al. (2019) presented a Digital Twin based IHD (Ischemic Heart Disease) detection architecture called "Cardio Twin". In the architecture, the authors categorized the myocardial and non-myocardial conditions with a CNN (Convolutional Neural Network). The employed CCN extracts features from the ECGs and determines the categorization. In the experimental study, the authors used a "PTB Diagnostic ECG Database" of Physio Bank. The study resulted in an accuracy of 85.77% and took 4.8 seconds for categorizing each sample. The demonstration proved that the technology is ready for adoption.
- Nonnemann et al. (2019) proposed "Health@Hand", an IoT based system in the digital twin of an ICU (intensive care unit) to monitor vital signs and administrative data. The system aims to summarize and process healthcare providers' health data in real-time. It provides a visual interface to the users for medical data analysis and management in different data sources.

The impact of DT in PHC with the integration of multiple technologies such as IoT, cloud, AI and BD is best expressed by a quote from Lucy to Charlie Brown in one of the comic strips of Peanuts: *Individually these fingers [in our case, technologies] may mean nothing, but put them together and you get a fighting force fearsome to behold.*

2.5 Synthesis and Key Take-Aways

Since the emergence of EHR in the healthcare industry, emerging technologies have been an indispensable part of healthcare management and care delivery. The recent increase in wearables usage for health fitness and the ongoing CoVID-19 situation has placed a timely demand for personalized digital health service to the industry. Based on the emerging information technologies discussed in this chapter, the healthcare industry has recently introduced different personalized digital health services. Precision Healthcare (PHC) is prominent among them.

There is, as yet, no universally accepted definition of PHC among scholars. Various synonyms and similar terms are found in recent studies (National Research Council, 2011; Collins & Varmus, 2015; Christensen et al., 2016; Gorini & Pravettoni, 2011; Happe & German Academy of Sciences Leopoldina, 2014; Hull, 2018; Teng et al., 2012; Huang & Dobbie, 2016; Colijn et al., 2017). Most of them are applied interchangeably, either representing the same concept or a wide range of ideas (Huang & Dobbie, 2016). Therefore, based on these empirical references, PHC can be defined as the incorporation of emerging information technologies to deliver customized healthcare service to individuals in optimized duration and cost applying navigation and scalability. By analyzing large health datasets using emerging information technologies such as IoT, Wearables, Cloud computing, Blockchain and smart contracts, Big Data, Artificial Intelligence, Machine Learning, Augmented and Virtual Reality, and Digital Twin, PHC purposes to enhance lifestyle, genomic and environment data understanding, and adjusts the findings in diagnosis, treatment and prevention (Kellogg et al., 2018). PHC utilizes the discovered data to deliver the right treatment at the right time (Hickey et al., 2019). It inspires patients for active participation in healthcare activities and health decisions (Hickey et al., 2019). PHC comprises potential of minimizing healthcare cost for each individual in a growing society.

2.5.1 PHC Implementations

PHC, including the use of genomic data, is still in the early discovery stage, even for early-adopting entities (Feero et al., 2018; *Center for Connected Medicine (CCM) and KLAS Research*, 2020). Numerous initiatives and researches are currently ongoing to harness its potential for next-generation healthcare. Table 2.2 provides a summary of significant PHC initiatives globally. It reveals the widespread undertaking of delivering precision health to achieve better outcomes.

2.5.2 PHC Initiatives from Commercial Entities

Apart from the state-funded initiatives, several promising commercial initiatives

48   *Capturing Value in Digital Health Eco-Systems*

Table 2.2. PHC Initiatives around the globe

Country	Initiative(s)	References
Australia	Australian Genomics Health Alliance	www.australiangenomics.org.au
	Australian Genomic Cancer Medicine Program	Australian Government Department of Health, 2018
	Genome Profiling Platform	Garvan Institute of Medical Research, 2018b
Canada	Genome Canada	www.genomecanada.ca
	GE3LS Network in Genomics and Personalized Health	Genome Canada, n.d.
Denmark	The National Biobank Register	www.nationalbiobank.dk
	Bio- and Genome Bank Denmark	www.regioner.dk/rbgben
	The Danish National Biobank at Statens Serum Institut	www.ssi.dk
	The Danish Neonatal Screening Biobank	www.ipsych.au.dk
	The Danish National Birth Cohort	www.ssi.dk
Estonia	The Estonian Biobank	www.biobank.ee
Finland	The National Institute of Health and Welfare (THL)	Njølstad et al., 2019
	FinnGen Project	
France	Genomic Medicine 2025	Alliance Aviesan, 2015
	National Network of Reference & Competence Clinical Centres for Rare Diseases	French National Plan for Rare Diseases 2018-2022, 2018
Germany	Personalised Medicine Action Plan	German Federal Ministry of Education and Research (BMBF), 2013
Iceland	deCODE genetics	Njølstad et al., 2019
Japan	AMED-GEM (Agency for Medical Research and Development-Genome Medical Alliance) project	AMED, 2019
	Program for an Integrated Database of Clinical and Genomic Information	
	Platform Program for Promotion of Genome Medicine	

(Contd.)

Table 2.2. (*Contd.*)

	BioBank Japan Project for Genomic and Clinical Research	
	Tohoku Medical Megabank Project	
New Zealand	Precision Driven Health (PDH) Program	Callaghan Innovation, n.d.
Norway	The Tromsø Study	www.tromsoundersokelsen.no
	The Norwegian Mother and Child Cohort Study (MoBa)	www.fhi.no/MoBa
	The Hordaland Health Studies	http://husk-en.b.uib.no
	The Health Survey of Northern Trøndelag (HUNT)	Njølstad et al., 2019
Singapore	Personalised OMIC Lattice for Advanced Research and Improving Stratification (POLARIS)	Agency for Science, Technology and Research, Singapore
Sweden	Genomic Aggregation Project in Sweden (GAPS)	GAPS, n.d.
	ANDIS	ANDIS, n.d.
Switzerland	The Swiss Personalized Health Network (SPHN)	Aronson & Rehm, 2015; Lawrence & Selter, 2017
	Swiss Institute of Bioinformatics (SIB)	Melier-Abt et al., 2018
	The BioMedIT project	
United Kingdom	The 100,000 Genomes Project	The 100,000 Genomes Project, 2018
	Personalised Medicine Strategy (England)	Personalized Medicine Strategy, 2015
	Scotland's Precision Medicine	Proposal of Recommendations for a Personalized Precision Medicine National Strategy, 2017
United States	Precision Medicine Initiative (PMI)	https://syndication.nih.gov/multimedia/pmi/infographics/pmi-infographic.pdf
	All of Us Research Program	https://allofus.nih.gov/
	Personalized Medicine Coalition (PMC)	Proposal of Recommendations for a Personalised Precision Medicine National Strategy, 2017

are also found in the studies. For instance, the US-based company, Illumina has developed "HiSeq X Ten", a supercomputer which is able to conduct 20,000 genome sequences (whole human) per year at an approximate cost of USD 1000 per genome. This is a thousand times cheaper than the cost in 2001 (*Business Wire*, 2014). Kaiser Permanente (an integrated managed care consortium from California) have adopted a PHC initiative to collect, study and utilize genomic and other healthcare data from their populations (Schaefer, 2011). Similarly, Geisinger (a regional health care provider to central, south-central and northeastern Pennsylvania and southern New Jersey) have taken the "MyCode" initiative to their service locations (*MyCode Community Health Initiative*, n.d.). Until January 2018, the number of participating Geisinger patients (with consent) was over 180,000, which is almost 90% of their patients (Carey et al., 2016; Williams et al., 2018). Foundation Medicine (a US company) is using PHC for cancer treatment (*Digital Transformation of Industries: Healthcare*, 2016). They determined two comprehensive genomic profiling tests that help oncologists to select the best-targeted treatment for patients. So far, they have tested around 60,000 patients with this approach (Foundation Medicine, 2015). Foundation Medicine is also collaborating with 20 pharmaceutical and biotech companies for research on new targeted therapy discoveries (*Digital Transformation of Industries: Healthcare*, 2016). Roche Diagnostics (a diagnostic division of Hoffmann-La Roche) and GE Healthcare (a US multinational conglomerate) have signed a long-term partnership agreement to design and develop a Clinical Decision Support System (CDSS) and Big Data analytics platform for PHC (Bresnick, 2018a). The partnership is focused on integrating EHRs, in-vivo and in-vitro data, real-time monitoring data, and clinical data to provide a tailored treatment to individuals.

2.5.3   Public-Private PHC Initiatives

Besides these, several public-private and academia-business initiatives are also found in the literature. For example, in 2015, the International Organization for Migration (IOM) organized a workshop on genomic-enabled healthcare learning (*Genomics-Enabled Learning Health Care Systems*, 2015). The workshop emphasized the learning cycle of genomic data analysis and how it impacts on clinical practices (Williams et al., 2018). Similarly, the University of California (San Francisco), the Bill and Melinda Gates Foundation, and the White House Office of Science, Technology, and Policy conjointly organized a PHC summit in 2016 (*Precision Medicine at UCSF*, 2016) to deal with the challenges and opportunities of PHC (Lyles et al., 2018). The National Institute of Health (NIH) is collaborating with eleven biopharmaceutical companies on the Precision Medicine Accelerating Cancer Therapies (PACT) program worth USD 215 million (Bresnick, 2017b). The collaboration is focusing on the identification and validation of biomarkers, which may lead to immunotherapies for a variety of cancer tests (Bresnick, 2018a). Memorial Sloan Kettering Cancer Centre, the Broad Institute of MIT, Harvard, Quest Diagnostics and IBM Watson Health are

An Overview of Emerging Digital Health Informatics

working together to provide a genomic-rich decision support system at the point of care (Bresnick, 2017a). Through the system, clinicians will be able to send biopsy reports of solid tumors to Quest Diagnostics. Later on the reports will be analyzed using IBM Watson's health genomic analysis tools to deliver treatment advice (Bresnick, 2018b). Indiana University (IU), the Regenstreif Institute (a US based research company) and LifeOmic (a technology company) are collaborating to develop "data commons" for PHC research with the intention of including clinical and genomic datasets (Bresnick, 2018b).

Most of the mentioned PHC initiatives are found government-funded. The majority of them are addressing population health (longitudinal data). Moreover, implementations are mostly disease-specific (oncology) (Health IT Briefing, 2020). Almost 70% of the initiates are either having lower patient engagement or non-existent (Center for Connected Medicine (CCM) and KLAS Research, 2020; Health IT Briefing, 2020). However, recent studies discussed in this chapter show potential of improving the current PHC context in Digital health. Adaption of these emerging information technologies into new digital healthcare streams such as PHC will be impactful towards the achievement of sustainable digital health for the future.

2.6 Chapter Recap

In the post-CoVID19 era, the application of digital health technologies will disrupt and transform conventional healthcare to personalized healthcare. Due to its nature of data dependency, personalised healthcare services' success entirely depends on the accuracy, efficiency, and specificity of the technologies used for connectivity, storage, security, data modelling, and analytics. Emerging digital technologies such as the types reviewed in this chapter represent the potential to contribute with innovative personalized healthcare services for Digital Health. This may well turn out to be the necessity that drives innovation in meeting the United Nations' Sustainable Development Goal (SDG 3), which advocates universal healthcare access. This chapter is by no means a definitive account but serves as an overview of the state-of-the-art of digital health technologies in the emerging health eco-system that is the focus of this book. It reveals the promise of emerging digital health technologies, especially in an eco-system where players are both informed as well as free to make informed choices.

CHAPTER

3

Review of Key Concepts and Methodology

Guided by the research problem and the research questions formulated in Chapter 1, this chapter builds on the systematic review of background concepts presented in Chapter 2. The reader is referred to Stephanie & Sharma (2019) for a deeper epochal review. The chapter also details the rationale adopted in selecting cases for in-depth study in phase 2 of this research.

Chapter 3 next describes the case study methodology adopted in this research. It begins with a rationalization of the choice of research methodology, outlines the research road map, and goes on to detail the specific steps followed in executing the three phases of this research. While phase 1 of this research was concerned with conceptualizing a patient-centric e-health ecosystem and the issues surrounding the materialization of such an ecosystem, phase 2 set out to test the conceptual model developed in phase 1 using case studies of unfolding national-level e-health initiatives such as Singapore's NEHR and the US' HITECH Program. Phase 3, on the other hand, had to do with a comparison of the insights drawn from each of the two case studies, for the purpose of deriving a common explanation that could help characterize the issues typically encountered in e-health implementations, and lead this research to identifying the critical success factors for e-health implementations. The chapter ends with a summary of key inferences and conclusions drawn from the discussion.

3.1 E-Health as Defined in Extant Research

The term e-health came to be used first in the year 2000 (Pagliari et al., 2005) and it still remains a grey area in view of the many varied definitions of its scope and focus (Pagliari et al., 2005; Mettler, Rohner & Baacke, 2008; Kivits, 2013; Treskes, Van Der Velde, Atsma & Schalij, 2016). According to Eysenbach (2001), for instance, e-health is the point of intersection where medical informatics, public health and business come together for a common purpose. In the view of Eysenbach (2001), it signifies health services and information delivered or enhanced through the Internet and related technologies. Pagliari et al. (2005)

Review of Key Concepts and Methodology **53**

draw attention to the fact that most e-health definitions place an emphasis on its communicative functions which are facilitated by networked digital technologies, especially the Internet. They therefore conclude that "e-health is the use of emerging information and communications technology, especially the Internet, to improve or enable health and healthcare" (p. 14). As different from these scholars, de Brantes, Emery, Overhage, Glaser & Marchibroda (2007) view e-health in terms of its potential to reduce or eliminate information asymmetries in the healthcare market place and transform it into a more transparent and efficient market. Busch (2008) points out that e-health involves digital and electronic tools and network exchanges conceived and developed to facilitate designated market players to generate, transfer and utilize healthcare data in electronic form. On the other hand, the World Health Organization (WHO) simply defines e-health as "the transfer of health resources and healthcare by electronic means" (WHO as cited in Treskes et al., 2016, p.443).

Despite such varied definitions, the common factor underlying e-health is that it comes with the promise of improved healthcare, reduced costs, reduced medical errors, increased efficiency of information flow and, most importantly, empowerment of healthcare consumers to take decisions on their own. This is the fundamental premise and promise of an e-health ecosystem.

3.2 Major Barriers to E-Health: A Game Theoretic View

The formal study of decision-making in strategic situations where several players must make choices that potentially impact the interests of the other players is called game theory (Turocy & von Stengel, 2001). The term "game" in game theory is used to formally describe a strategic situation. Game theory is a set of analytical tools that helps comprehend the dynamics that unfold when decision-makers interact (Osborne & Rubinstein, 1994). It provides a systematic way to understand the behavior of players in situations of interdependent fortunes (Brandenburger & Nalebuff, 1995) like in the e-health ecosystem. It is the study of conflict and cooperation among intelligent, rational entities often referred to as 'players', in their decision-making processes (Myerson, 1991). Since e-health calls for cooperation among the players amidst conflicts, it is believed that game theory principles may be useful for an analysis of their behavior. In fact game theoretic frameworks have been utilized by scholars like Bandyopadhyay, Ozdemir & Barron (2012) and Martinez, Feijoo, Zayas-Castro, Levin & Das (2016) to investigate issues in e-health in other contexts. While the former group used game theory principles to investigate if PHRs had the potential to spur EHR adoption among the healthcare providers, the latter group developed a game theoretic model to predict the willingness of healthcare providers to exchange patient information with other unaffiliated healthcare organizations including competitors.

A basic aspect of a game is the interdependence of the players' decisions (Dixit & Nalebuff, 2008), which, in the context of e-health, may mean a key player's freedom of choice whether or not to participate and create values in the ecosystem. If the healthcare providers for example are not motivated to participate in e-health and create values in the form of EHRs, significant values may be lost for the ecosystem rendering it thereby unsustainable. For every key player to participate and create value in the e-health ecosystem what is basically needed is fairness in terms of the values that can be captured from the ecosystem.

3.2.1 Fairness

The notion of fairness has to do with ensuring that all players in a game get a fair deal in their interactions, agreements or situations. According to Grandori (1999), fairness denotes rules or criteria used to divide valuable resources for apportionment among different players. Off the four rules the concept of fairness suggests, the input-output rule is obviously the most relevant for the e-health ecosystem. The input-output rule states that fairness is the correspondence between the pay-off received by a player and their contribution to the achievement of the total output. In the context of e-health, this rule would mean that every player should be able to capture values in proportion to the value they create in the network. Such fairness is essential to keep the players engaged with the ecosystem, without which the benefits of e-health cannot be harnessed in full.

3.2.2 Efficiency

It may be noted that the notion of efficiency may conflict with the notion of fairness in the digital industry which is characterized by intense competition resulting from reduced information asymmetries. In healthcare, resource misallocations are massive owing to the lack of uniform and transparent information in the marketplace (Shmanske, 1996; Yip & Hsiao, 2009). Some key players have remarkable incentives to hoard information and leverage the asymmetry to maximize their profits. A case in point is the healthcare providers who have more incentive to institutionalize patient data than to share it with other players. The consequence is a serious lack of coordinated care, the brunt of which has to be borne by healthcare consumers (de Brantes et al., 2007). No doubt, the e-health ecosystem can significantly reduce such asymmetries and improve market transparency and efficiency, but it may not provide the financial stimulus for such players to participate in the ecosystem. The ecosystem would be sustainable only if the right tradeoff between efficiency and fairness is achieved. Another possibility worth investigating is whether the ecosystem may still be sustainable if at the least, no player is worse off than status quo because of their participation in the ecosystem, and at least one player (healthcare consumers) is better off. Such an outcome is referred to as 'Pareto-efficient'.

3.2.3 Pareto Efficiency

A possible outcome of a game is Pareto-efficiency which implies that there is no other outcome that makes every player at least as well off and at least one player strictly better off. That is, a Pareto-efficient outcome cannot be improved upon without hurting at least one player (Grandy, 2006). However, whether this notion would hold good as a basic criterion for the sustainability of e-health, remains to be investigated.

A game theoretic view of the major barriers to e-health particularly from the healthcare providers' perspective makes it evident that these are manifestations of the prisoner's dilemma, a classic example to demonstrate game theory. Prisoner's dilemma is a game theoretical model of cooperation and conflict, originally developed by Merrill Flood and Melvin Dresher in 1950 and formalized and nicknamed by Albert W. Tucker a little later (Surhone, Timpledon & Marseken, 2010). Luce and Raiffa (1957) in their book, *Games and Decisions*, describe Prisoner's Dilemma as follows:

Two suspects are taken into custody and separated. The district attorney is certain they are guilty of a specific crime, but he does not have adequate evidence to convict them at trial. He points out to each prisoner that each has two alternatives: to confess to the crime the police are sure they have done, or not to confess. If they both do not confess, then the district attorney states he will book them on some very minor trumped-up charge such as petty larceny and illegal possession of a weapon, and they will both receive a minor punishment; if they both confess they will be prosecuted, but he will recommend less than the most severe sentence; however, if one confesses and the other does not, then the confessor will receive lenient treatment for turning state's evidence whereas the latter will get "the book" slapped on him. (p. 95)

Prisoner's dilemma, in other words, is the conflict between self-interest and group interest. Rapoport and Chammah (1965) define prisoner's dilemma as a mixture of interpersonal and intrapersonal conflict, which eventually leads to individual defections culminating in an overall scenario of less desirable outcomes (Turocy & von Stengel, 2001). In the context of e-health, these are some typical dilemmas faced by healthcare providers during the various stages of the evolution of e-health and may be considered to fall into two categories namely (i) participation dilemmas and (ii) cooperation dilemmas. These dilemmas, if unresolved, may render e-health infeasible and unsustainable. A brief account of each of the dilemmas is given below.

3.2.4 Participation Dilemmas

These are barriers that deter healthcare providers from taking the essential first steps towards e-health, which involves making substantial investments in building EHRs. What follows is a short discussion of the participation dilemmas:

3.2.4.1 Productivity Paradox

The famous quip by Robert Solow, Nobel Laureate in Economics, that, "we see computers everywhere except in the productivity statistics" (Solow, 1987), still rings true after decades, especially in the context of investments in e-health. Some healthcare organizations, for instance, still challenge the much-advocated link between investments in technology and improved organizational performance, keeping alive the debate on IT payoff, referred to in literature as the "productivity paradox" (Brynjolfsson & Hitt, 1998; Devaraj & Kohli, 2000).

In the absence of demonstrable evidence of positive payoffs from e-health investments (Wiedemann, 2012; Bergmo, 2015), the strategy most prevalent among individual healthcare providers is defection to e-health rather than joint cooperation. This would mean that if healthcare providers are not motivated to invest in EHRs due to the productivity paradox, e-health may continue to remain a distant dream.

3.2.4.2 Tragedy of the Digital Commons

EHRs are the building blocks of e-health that need to be heavily invested in and created by the healthcare providers. These digital health records also need to be enabled for exchange and reuse of health data by other players in the network such as patients, payers, vendors, and other healthcare providers (Adler-Milstein & Bates, 2010), so that the benefits of e-health are harnessed in toto. In other words, in a patient-centric e-health ecosystem, health data is viewed and treated as a public good or "commons" which every stakeholder including patients and those authorized by patients can consume without necessarily contributing to it. Albanese and Fleet (1985) describe this as 'free-riding' where a member of a group benefits from its access to a common resource more than they actually contribute to the cost of this common resource. This 'free-riding' is reminiscent of the misaligned incentives discussed in the context of e-health and often deters healthcare providers from investing in e-health (Bandyopadhyay et al., 2012) which in turn might prove detrimental to other players. If healthcare providers shirk from investing in e-health, it will only lead to a deficient or less desirable outcome for everyone, resulting in a situation referred to by Adar and Huberman (2000) as the 'tragedy of the digital commons'.

3.2.5 Cooperation Dilemmas

These dilemmas, though different from participation dilemmas, relate to such healthcare providers as have already invested in EHRs for productivity gains, but are reluctant to share the EHRs with other players in the network. Two such dilemmas are:

3.2.5.1 Information Asymmetry

It is well-acknowledged that the physician-patient relationship is characterized by asymmetric information (Arrow, 1963; Blomqvist, 1991). This is because a

Review of Key Concepts and Methodology

physician who examines a patient acquires information about the patient which the latter cannot access on his/her own (Blomqvist, 1991). Such information asymmetry results in provider-centrism where the providers are very much in control of their patients' healthcare decisions and choices, which may not always be in the patients' interest.

However, given the potential of e-health to foster a sharing and exchange of health data, it has been suggested from time to time that information asymmetries of this kind may be reduced to facilitate greater patient empowerment.

Though the health data of individual patients is maintained by healthcare providers in heavily invested systems, it is by all means a common property resource owned by both healthcare providers and patients and it should therefore be made accessible to both groups. However, the reality is that such health data is underutilized, if not totally unutilized, especially by patients for reasons beyond their control (Martinez et al., 2016). One basic reason is that healthcare providers who have invested in EHRs are unwilling to progress to the next level by sharing the data with their patients and other players in the network. If such information sharing is made possible as in an efficient market system, patients would really be empowered to shop around and choose a healthcare provider on their own, based on criteria such as cost-effectiveness, reliability and quality (Hill & Powell, 2009). This would go a long way in reducing provider-centrism, as well as, providers' return on EHR investments.

3.2.5.2 Information Blocking

Some healthcare providers may be willing to invest in EHRs if they perceive certain significant productivity gains from the investment. Furthermore, they may even take e-health to the next level by sharing their patients' health data with parties outside their institutional walls. However, they may limit such sharing to a select group of partners within the system or network 'to maintain a captive market share and reinforce market dominance' (Martinez et al., 2016, p. 2). This may endanger the interoperability, and hence, the exchangeability and reusability of health data beyond the network. This results in patients' choices getting restricted to a few partners or players carefully selected by the healthcare provider in extreme self-interest. Such an outcome defeats the very purpose of e-health namely patient-centrism which means unlimited and ubiquitous access for patients to their health data.

It is against such a backdrop that this study aims to investigate the sustainability of a patient-centric e-health ecosystem with particular focus on the dilemmas of the healthcare providers without whose participation and cooperation patient-centric e-health may not be realized. In other words, this study aims to unlock the latent value of e-health technologies by discovering a logic that can align the technical potential of e-health with realization of benefits for all its stakeholders. This logic or 'heuristic logic' is what is referred to as a **business model** (Chesbrough & Rosenbloom, 2002).

3.2.6 Business Models in E-Health

This section begins with a short discussion on the concepts of business model and value notions, and then proceeds to present a synmonograph of the various e-health business models that were observed through thematic reviews of related literature and industry trends. The reviews concerned were conducted with a view to comprehending the workings of the various types of e-health business models that are in existence.

A business model, simply defined, is the representation of a firm's core logic for creating and capturing value in a value network (Shafer, Smith & Linder, 2005; Spil & Kijl, 2009), and is geared towards creating value for all the parties involved (Zott & Amit, 2010). It is the blueprint for how a business can be organized for the benefit of all the stakeholders. Value is 'created' when a firm (player) develops its core competencies, capabilities and advantages to perform work activities that differentiate it from competitors. Value is said to be 'captured' when the firm derives economic returns in relation to the value it creates (Shafer et al., 2005). A comparison of the values created and captured in the e-health network will help evolve a business model to promote 'collaboration for value' which will lead to a fair, efficient and sustainable e-health ecosystem.

Zott & Amit (2010) visualize a firm's business model as transcending the boundaries of the firm on account of the interdependent activities it engages in with its partners. As such, e-health business models may take on several forms, each with a different purpose and having its own revenue and cost structure (Parente, 2000). From the standpoint of e-commerce relationships among the various players, the most popular forms identified are portals, connectivity, business-to-business (B2B), and business-to-consumer (B2C) (Parente, 2000; Payton, 2003). It may be noted that the emergence of other forms such as consumer-to-consumer (C2C) (Eysenbach, 2001; Broderick & Smaltz, 2003) was predicted more than 10 years ago, and some of them have already been in existence for some time now (e.g. www.ivillage.com).

Another perspective of e-health business models is their orientation or "centrism". Although all models operate with the common goal of improving patient care, some models are provider-centric, primarily focusing on designing and creating interoperable EHRs integrated with medical imaging, therapy, and laboratory diagnostics. Others are patient-centric and focus on personal health records (PHRs) thereby putting patients in control of their health information. The onset of the latter trend was documented as early as 1999 (Kilgore, 1999), with the past decade seeing a gradual shift in focus from provider-centric applications to consumer-centric models that empower healthcare consumers by allowing them to access, manipulate and understand data about their personal health (Burkhard, 2009; Zieth et al., 2014). Such a shift is necessitated by a growing awareness of patients and others, of the need to take control of their health (Purcarea, 2009, Truog, 2012). In support of this trend, findings from several studies claim that a

Review of Key Concepts and Methodology **59**

large majority of the healthcare consumers consider PHRs useful and evince an interest in building their own PHRs (Peters, Niebling, Slimmer, Green, Webb & Schumacher, 2009; Ford, Hesse & Huerta, 2016). Regardless of centrism, some business models are proprietary (e.g. Kaiser Permanente's My Health Manager), while others are open-source (e.g. Microsoft Health Vault).

A look at the e-health market trends may also reveal the existence of other interesting infomediary business models which are rather specialized. Verisk Health[1] recently rebranded to Verscend Technologies (Business Wire, 2016), for instance, has gone through sweeping changes in its business model over the last few years. From a company specializing in providing medical record retrieval services to providers, insurers, patients and law firms on a pay-per-record basis, it has now moved towards data-driven healthcare solutions. Keas[2] is another infomediary, that targets major self-insured employers to market its employee wellness program, a health management platform, aimed at reducing these companies' healthcare costs. A further example is IBM[3] which focuses on health analytics solutions to cater to the growing demand for leveraging patient data to extract business insights.

Although a variety of models are used by e-health infomediaries to organize their business, their economic survival has often been questioned because the payoffs they can extract from the network are closely tied to the benefits their co-players can extract from the network (Ford et al., 2004; Konrad & Peter, 2007). A case in point is Google Health which decommissioned its PHR product in early 2012 after dabbling with it for five years (Spil & Klein, 2014). Based on a scan of the current e-health market trends, it is found that revenues for an e-health infomediary may be generated through various sources ranging from licensing fee (e.g. Microsoft Health Vault), to sale of wellness programs (e.g. Keas), to sale of analytics solutions (e.g. IBM) and transaction fees generated by B2C, B2B and / or C2C ecommerce activities on its network (e.g. VeriskHealth, Keas, iVillage). The scenarios for each of the possible revenue sources for an e-health infomediary are summarized in Table 3.1.

3.3 Vocabulary of Research Concepts

It is hoped that the literature review included in this chapter brings together the vocabulary that needs to be assimilated in order to be able to develop a sound understanding of the phenomenon under study namely, e-health, and its impact on the healthcare industry. Table 3.2 provides a subset of the vocabulary and their meanings.

---

[1] http://www.veriskhealth.com/
[2] www.keas.com
[3] http://www-01.ibm.com/software/analytics/healthcare/solutions.html

60   *Capturing Value in Digital Health Eco-Systems*

Table 3.1. Revenue sources for E-Health infomediaries

Revenue source	A typical scenario
Advertisements	The e-health firm may sell web inventory to e-health vendors who wish to promote their products or services to the e-health network members through advertisements e.g. it may charge the vendors on a Cost Per Mille (CPM) basis.
Subscription Fees	The e-health firm may offer premium services to special interest groups on a subscription basis e.g. it may provide updates on new medical technologies to surgeons for a yearly subscription fee.
Transaction Fees	The e-health firm may exact transaction fees when it moves data over the Internet from one network member to another e.g. it may move a patient's medical history from a physician to a pharmacy for a transaction fee.
Business-to-Consumer Ecommerce	The e-health firm may facilitate sale of healthcare products or services by a business on its network to consumers on its network for a commission e.g. it may earn a commission when a patient (consumer) purchases medicines from a pharmacy (business). Additionally, it may also charge the business a listing fee for inclusion in its network.
Business-to-Business Ecommerce	The e-health firm may facilitate sale of healthcare products or services by a business on its network to another business on its network for a commission e.g. it may earn a commission when a hospital (business) purchases refurbished medical equipment from another hospital (business) through online auctions. Additionally, it may also charge the businesses a listing fee for inclusion in its network.
Consumer-to-Consumer Ecommerce	The e-health firm may facilitate sale of healthcare products by a consumer on its network to another consumer on its network for a commission e.g. it may earn a commission when a healthcare consumer purchases a used treadmill from another healthcare consumer through online auctions.

3.4 Selection of Cases for Phase 2

This section discusses the sampling strategy adopted to select specific cases for study during phase 2 of this research. The approach adopted was 'purposive sampling', whereby the cases were selected based on the following sampling strategies:

Maximum variation – According to Patton (1990), this strategy refers to selecting cases that have diverse variations, but at the same time, offer potential to strengthen the results. This is because, any common patterns or explanations that emerge from very different cases, can help capture "core experiences and central, shared

Review of Key Concepts and Methodology **61**

Table 3.2. Vocabulary of research concepts

Concept	Definition
E-Health	The use of emerging information and communications technology, especially the Internet, to improve or enable health and healthcare.
Interoperability	The ability of two or more IT systems to exchange and use information. Interoperability is a prerequisite for binding together a huge network of critical and real-time health data so as to benefit from e-health technologies.
Electronic Medical Records (EMR)	Computerized clinical records of an individual created disparately by healthcare providers such as hospitals and physician offices. They do not enable sharing of health information because they do not conform to nationally recognized interoperability standards.
Electronic Health Records (EHR)	A longitudinal electronic record of an individual's health information that conforms to nationally recognized interoperability standards and, therefore, has the ability to follow the individual through different modalities of care from various healthcare providers. It also enables sharing of health information among authorized stakeholders.
Personal Health Records (PHR)	An electronic record containing health-related information that is managed, shared, and controlled by an individual, and conforms to nationally recognized interoperability.
Provider-centric	Refers to healthcare IT systems that are focused on benefits for healthcare providers. For example, the EHR is a provider-centric system that is primarily intended to support healthcare providers' clinical decisions.
Patient-centric	Refers to healthcare IT systems that are focused on benefits for patients or healthcare consumers. For example, the PHR is a patient-centric system that is primarily intended to support patient or consumer empowerment.
Cloud computing	The technology of delivering computing as a utility. It refers to a range of IT services delivered over the Internet - Software as a service (SaaS), Platform as a service (PaaS) and Infrastructure as a service.
Infomediary	A neutral, unbiased, third-party entity that functions as a digital information conduit and business matchmaker.
Game theory	The study of conflict and cooperation among intelligent, rational individuals or entities in their decision-making processes.
Fairness	The correspondence between the pay-off received by a player and their contribution to the achievement of the total output.
Pareto-efficiency	An outcome of a 'game' that makes every player at least as well off and at least one player strictly better off.

(Contd.)

Table 3.2. (*Contd.*)

Concept	Definition
Prisoner's dilemma	Refers to the conflict between self-interest and group interest.
Free-riding	Refers to a situation where a member of a group benefits more from its access to a common resource, than it actually contributes to the cost of this common resource.
Tragedy of the commons	Denotes a situation where the rational behavior of an intelligent individual or entity will be to exploit a freely available common resource for self-gain rather than consider the best interest of the whole group. In the digital world such behavior leads to a deterioration of the system making everyone worse off and resulting in what is known as the tragedy of the digital commons.
Information asymmetry	Refers to a situation where information is not uniformly and transparently available to the various parties involved in a transaction.
Information blocking	Denotes a situation when persons or entities knowingly and unreasonably interfere with the exchange and use of electronic health information.
Value creation	Occurs when a firm (player) develops its core competencies, capabilities and advantages to perform work activities that differentiate it from competitors.
Value capture	Occurs when the firm derives economic returns in relation to the value it creates.
Business model	A description of the logic using which a firm will create and capture value in a business ecosystem.

aspects" (Patton, 1990, p. 172). Since this research aims to identify the critical success factors for implementing patient-centric e-health, it was believed that this strategy could help improve the analytic generalizability of the findings, which would be based on common patterns that cut across the variations in the cases. By applying this strategy, the cases that were selected for study were Singapore's NEHR and the US' HITECH Program. While the former is a high performing healthcare system, the latter is a low performing healthcare system. Needless to say, it would be of significant interest if some commonalities can be deduced across these diverse cases that would contribute to identifying the critical success factors for e-health implementations.

Information richness – Miles and Huberman (1994, p. 34) contend that it is crucial for the sampled cases to have the potential to provide rich information on the phenomenon of interest. The cases that met this criterion were again Singapore's NEHR and the US' HITECH Program. During the process of reviewing literature and industry reports to identify potential cases for study as well as, while soliciting deeper insights into these cases from identified representatives, it became evident

Review of Key Concepts and Methodology 63

that these were the only cases which offered scope for an intensive study. This was because, not only was a variety of secondary data sources available for the two cases, but also, access to primary data was possible in the form of interviews that materialized with some identified representatives of these cases. It was firmly believed that the insider knowledge elicited from these representatives would complement the knowledge gained through literature and document reviews, to produce information-rich cases. Moreover, the two e-health initiatives selected for study have made significant progress relative to most others which are still in the concept or early implementation stages. This was yet another characteristic which was believed to further enhance the information-richness of the two cases selected for study.

3.5 Choice of Research Methodology: Rationale

The case study strategy enables the researcher to obtain a holistic view of a complex phenomenon (Gummesson, 1991) by tapping into multiple sources of evidence. And, it is often the preferred strategy when the focus of the research is on a contemporary phenomenon within real-life context (Yin, 2003, p. 1). Moreover, the nascence and the richness of the field of e-health make it a good candidate for the case study approach which is believed to help understand the dynamics and intricacies involved in the domain. Specifically, the focus of the current study, namely e-health ecosystems, is a topic of interest that is inarguably complex and abstract, and therefore warrants the use of the case study approach to ensure that the topic is well explored.

Eisenhardt (1989) describes case study as an approach that can involve single or multiple cases, and as an efficacious strategy to understand the dynamics within a single setting. There are no fewer than three different types of case study approaches depending on the purpose of the study – exploratory, explanatory or descriptive (Yin, 2003). Exploratory case studies are undertaken to investigate a new field of research perhaps with a view to formulating an appropriate research design and hypo-monograph; explanatory case studies seek to study complex causal links and descriptive case studies attempt to describe specific characteristics of a phenomenon in its natural setting. It needs to be stated here that the context of the current study lends itself well to a descriptive, multi-case study.

3.6 The Research Approach

The roadmap followed for conducting this research is depicted in Figure 3.1.

Phase 1 involved developing an initial theoretical framework backed by literature search. This theoretical framework helped identify key concepts and the relations among them which were believed to account for the phenomenon investigated, namely e-health. It further provided a structure to conceptualize a patient-centric e-health ecosystem – the key players, the potential digital data flows and the values thereby created and captured by the key players. Such

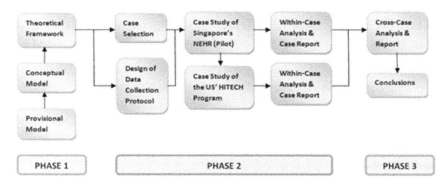

Figure 3.1. Research roadmap

conceptualization was deemed necessary in the light of the research objective to educe the critical success factors for a patient-centric e-health ecosystem. The conceptualization was accomplished by employing multiple sources of data such as thematic literature reviews, observations and expert interviews. Phase 1 culminated in the development of a conceptual model of a patient-centric e-health ecosystem which also provided answers to the set of questions raised under RQ1. Phase 1 is covered in Chapter 4.

Phase 2 focused on testing the conceptual model developed in phase 1. The testing was intended to examine the extent to which the conceptual model reflected reality as well as to investigate how the notions of fairness and efficiency should be traded off for achieving a sustainable e-health ecosystem. Towards this end, in-depth case studies of two national-level e-health ecosystems – Singapore's NEHR and the US' HITECH program, were undertaken. The within-case reports for the NEHR and HITECH program are presented in Chapter 5 and Chapter 6 respectively.

Phase 3 was concerned with extracting insights from the two case studies undertaken in phase 2 for the purpose of characterizing the issues surrounding adoption of e-health at a national level as well as identifying the critical success factors needed for establishing a national level e-health ecosystem. This was achieved through a cross-case analysis where the insights gleaned from the individual case studies were compared so as to derive common patterns and explanations which would in turn provide answers to RQ2. Phase 3 is covered in Chapter 7.

The following sections are intended to describe the specific steps involved in executing the three phases described above.

3.7 Phase 1: Development of Conceptual Model of a Patient-Centric E-Health Ecosystem

This phase of the research was aimed at developing a conceptual model of a

Review of Key Concepts and Methodology **65**

patient-centric e-health ecosystem by utilizing three sources of data, namely thematic literature reviews, industry observations and expert interviews.

The expectation was that the data thus collected would help ascertain the key players in a patient-centric e-health ecosystem, the digital data flows that are made possible in such an ecosystem, and the values that can consequently be contributed to and captured from the ecosystem by the key players. It was imperative to develop a sound understanding of these aspects of the e-health ecosystem, so that a qualitative game theoretic analysis could subsequently be undertaken to determine if the values created and captured by each player participating in this ecosystem are equitable enough to warrant their participation. This would in turn help determine whether a patient-centric e-health ecosystem, the way it is described in this research, is feasible and sustainable. Besides, it would also help focus on what needs to be done to make such an ecosystem a reality. Given below is a detailed description of the sources of data discussed previously.

3.7.1 Thematic Literature Review

A highly focused review of relevant literature is known as a thematic literature review. Urquhart and Fernandez (2013) observe that the theoretical data collected through such a review not only enables researchers to exploit extant literature to advantage, but also enriches their study by facilitating comparison with emergent themes from data obtained through the other sources.

In this research, such a literature review was done using the following steps:

1. A search for journal articles relevant to the substantive areas of this research was conducted using Google Scholar. Keywords such as 'e-health', 'electronic health records', 'electronic medical records', 'personal health records', 'game theory' and 'business model' were used to identify journal articles relevant to the 'substantive areas'.

2. Once a journal article was identified to be relevant on the basis of its abstract, its full-length pdf was downloaded from the database for an in-depth review. Examples of such databases used in this research are Academic Search Premier, Business Source Premier, ScienceDirect, MEDLINE Complete, ABI/INFORM and ProQuest.

3. Journals carrying relevant articles were earmarked for the purpose of further exploration of sources for pertinent literature. Health Affairs, Healthcare Financial Management, Health Management Technology, Physician, Communications of the ACM, Harvard Business Review, and British Medical Journal were but a few of the many journals referred to in this research.

4. To supplement the data from the academic journals, iterative searches of internet-based resources such as white papers and conference papers were also performed, and more data sources identified.

5. As and when data was extracted from these searches, it was simultaneously analyzed by comparing it with existing data for convergence or divergence, as this would help guide future data collection efforts.

3.7.2 Industry Observations

In addition to having a thematic literature review, it was also necessary to keep track of emerging trends in the industry. This was considered even more crucial in the context of e-health which keeps evolving astonishingly rapidly. Observing and analyzing such trends and developments alongside the data from the literature review would help construct a more current and, at the same time, a more complete model of e-health. This step was sought to be accomplished through iterative searches for Internet-based sources such as industry reports, online magazines, news websites and organizational websites.

In the manner outlined above, more than two hundred journal articles, online resources and industry reports were perused. The e-health model thus developed based on a thematic literature review and observations of current industry trends would however be considered only provisional; its robustness could be ascertained only with reference to yet another source of data, namely expert interviews.

3.7.3 Expert Interviews

It was also considered necessary to develop a research instrument for in-depth interviews with industry experts in order to ascertain the robustness of the provisional model before embarking on further research. The objective of these expert interviews was two-fold: to validate the provisional model of the e-health ecosystem derived through a thematic literature review and industry observations, in the first instance, and to elicit expert opinion on the game theoretic notions of fairness and efficiency in the e-health ecosystem. This would help determine if different experts concurred with literature on the perceived imbalance in the existing business model arrangements among the key players. Section 3.2, 'Major Barriers to E-health: A Game Theoretic View' presents the conceptual framework used to investigate the correspondence between the values created and captured in the e-health ecosystem. Besides, the section provides an overview of these notions as well.

3.7.4 Research Instrument

The research instrument to interview industry experts was a structured interview template that comprised two sections (Appendix A). The conceptual model of digital flows and the research instrument were first peer-reviewed at the 2nd Symposium on Healthcare Advances in Research and Practice (SHARP 2.0), held during April 28-29, 2011, in Moorhead, Minnesota (USA). Next, as a pilot run, the process was also repeated at the UPMC (University of Pittsburg Medical Centre) in Pittsburg, Pennsylvania (USA) on May 2, 2011. The feedback obtained

Review of Key Concepts and Methodology **67**

through these pilot tests was used to further refine the questionnaire with a view to establishing its validity and reliability. The final questionnaire included eleven questions which were distributed over two sections. A brief account of each of the sections is given below.

3.7.4.1 E-Health Ecosystem

This section provided an illustration of the provisional model, with some background on the basis upon which this model was derived. The informants were required to go through this background and give their view on whether the model adequately reflected reality. They were also expected to point out omissions or redundancies, if any, in the model, stating valid reasons.

3.7.4.2 Values-Created vs. Values-Captured by Primary E-Health Players

This section listed the sources of values-created versus values-captured that were identified for each key player in the e-health ecosystem. It provided some theoretical background of game theoretic value notions to establish an understanding of these concepts in the context of e-health. The informants were asked to discuss whether these values were adequately identified and represented for each of the key players, and to point out discrepancies, if any, stating reasons. Opinions were also sought on: whether the value captured was greater or less than the value created for any player(s); which player(s) contributed (created) the greatest and least values to the ecosystem; which player(s) derived (captured) the greatest and least values from the ecosystem; the notions of fairness and efficiency as necessary conditions for the sustainability of the e-health ecosystem.

3.7.4.3 Target Informants

Twenty-one professionals across India, Malaysia, Singapore and the United States, representing different sectors of healthcare organizations such as Ministries of Health, IT departments of healthcare providers, health IT vendors, and healthcare BPOs were identified for expert interviews. A summary of the target informants' profile is presented in Table 3.3. It was ascertained that the potential informants were involved in the phenomenon of e-health in some capacity. While some of these target respondents were from professional circles or referrals through professional networks, the others were identified through LinkedIn, a social media platform. The tool used for selecting the informants was 'purposive sampling' which means deliberate selection of informants on the basis of qualities possessed by them such as knowledge and experience required to inform this research. Most importantly it was expected that the informants had both practical insider knowledge and willingness to act as surrogates for a wider circle of players in the e-health ecosystem. The informants who contributed to the study are highlighted in bold font. It may be observed that these informants comprised a good mix of senior and middle management professionals.

Table 3.3. Phase 1: List of target informants

Informant	Position	Organization
#1	Business Manager, Healthcare Informatics (Global)	**Crimsonlogic, Singapore**
#2	Senior Manager (Quality)	Surescripts LLC, USA
#3	Program Manager	US Dept. of Veteran Affairs, USA
#4	**IT Manager**	**Tan Tock Seng Hospital, Singapore**
#5	**Director of Industry Research, Institute for Communication Technology Management**	**The Marshall School of Business, University of Southern California, USA**
#6	**Senior Consultant, Health Services Group**	**Ministry of Health, Singapore**
#7	**Owner**	**ManyMedia, USA**
#8	**Assistant Director (Information Systems Division)**	**MOH Holdings Pte Ltd, Singapore**
#9	Vice President, Technology and Innovation	Western Interstate Commission for Higher Education, USA
#10	Vice President (Client Relations)	E4E Healthcare Business Services, India
#11	**Vice President (Sales)**	**E4E Healthcare Business Services, India**
#12	Business Development Director	CMPMedica, Singapore
#13	**Project Manager (Healthcare IT Systems)**	**Friar Tuck, Singapore**
#14	Associate Professor, Centre for Computational Intelligence	Nanyang Technological University, Singapore
#15	Senior Healthcare Consultant	Frost & Sullivan, Malaysia
#16	**Database Architect, Virtual Clinical Image Management**	**Accelerad, USA**
#17	Principal, iPMO	MOH Holdings Pte Ltd, Singapore
#18	**Manager (Clinical Governance)**	**Singapore General Hospital, Singapore**
#19	**Research Manager (Government Insights & Health Insights, Singapore)**	**IDC, Singapore**
#20	Manager (Healthcare Practice APAC)	Frost & Sullivan, Singapore
#21	**Founder and Managing Director**	**Venture E (Shanghai) Co., Ltd**

Review of Key Concepts and Methodology 69

3.7.4.4 Methodology

The twenty-one target respondents were sent an introductory email along with the interview template in early June 2010, inviting them to participate in the expert interview via email. They were requested to respond within a week, failing which they were sent a reminder. To boost the response rate, they were given the options of phone interviews and face-to-face interviews. A week after the first reminder, a second reminder was sent to those who had not responded by that time.

Some respondents who were available for a face-to-face interview were met personally to get the interviews done. A few others sought clarifications over the phone before completing the interview and emailing it back.

3.7.4.5 Response Rate

As of 30 September 2010, the final deadline, the response rate was 57% with twelve out of the twenty one targeted experts responding to the interview.

3.7.4.6 Analysis

The analysis was primarily aimed at validating the key players in the e-health ecosystem modelled in this research, confirming the values created and captured by these players as identified through thematic literature reviews. It was also intended to solicit opinions on whether there was correspondence between the values created and the values captured in respect of every key player (whether a player created more value than they captured, and vice versa).

To analyze the interview transcripts, coding approaches suggested by Corbin and Strauss (1990), Burnard (1991), Charmaz (2006) and Saldana (2009) were explored and evaluated systematically before a decision was made to go ahead with the techniques advocated by Saldana (2009). Such a decision was primarily made on account of the fact that Saldana's work presented a 'pragmatist approach' intended to help researchers select "the right tool for the right job" (p. 2) depending on the context. Besides, the approach was objective in that it was free from any bias for or against a specific research genre or methodology.

The coding scheme is presented in Appendix B. Expert responses to each interview question were addressed individually and coded in a manner that captured the essence of the response. Depending on the complexity of the response, multiple codes were sometimes identified for a single response. For each of the eleven questions, the codes were analyzed to uncover the presence of any patterns. Characteristics of patterns like frequency (happen often or seldom), similarity (happen the same way), correspondence (happen in relation to other events) and rarely, causation (one causing another), were observed. For the purpose of assessing **coding reliability** two coders including the first author were involved in the coding process. The codes were finalized once an adequate level of agreement was reached between the coders.

3.7.4.7 Findings

The findings based on the analysis of the qualitative data received from the twelve experts were compared with those from the thematic literature reviews and industry observations, and inferences were drawn. These expert interviews provided invaluable inputs which could help define the provisional models derived through thematic literature reviews and industry observations.

Chapter 4 presents the data, analysis and findings from phase 1 of this research.

3.8 Phase 2: Validation of Conceptual Model in the Field

One of the biggest issues confronting e-health is a lack of demonstrable evidence of its benefits for its investors as well as its sustainability over a period of time, which this phase of the research aimed to probe and address using the case study approach. Phase 2 of this research was thus concerned with field testing the conceptual model derived in phase 1 using case studies. The test was intended to determine the extent to which the sampled e-health ecosystems conformed to the conceptual model. It was of particular interest to investigate the dilemmas confronting the healthcare providers as well as the notions of fairness and efficiency in these ecosystems (discussed in section 3.2, 'Major Barriers to E-health: A Game Theoretic View') with a view to evolving some critical success factors which would bolster sustainability of the ecosystem. Also, on the agenda for this phase of the research was finding an answer to the question whether an e-health ecosystem on a cloud (health cloud) would prove to be one such design characteristic.

3.8.1 Sources of Data for Phase 2

As e-health is a relatively new and evolving field, it was hoped that this study would benefit greatly from research already undertaken in this area. This would then be complemented with primary data gathered through qualitative research interviews. These two slices of data would be utilized for the purpose of conducting case studies of seemingly successful e-health implementations around the world. The conceptual model developed in phase 1 of this research would be tested through in-depth case studies which would in turn utilize the slices of data collected through secondary research and expert interviews. It was hoped that a comprehensive approach such as this would help investigate the perceived imbalance in values-created vs. values-captured for the various players in the e-health network, and lead this research on to the process of deriving the critical success factors of a sustainable ecosystem. A brief discussion on the importance and relevance of the data sources for the case studies follows.

3.8.1.1 Secondary Research

The qualitative case study approach uses a variety of data sources as evidence, some of which can be accessed through secondary research. An obvious advantage of secondary research may be economy. Additionally, it provides access to information that would normally be inaccessible through primary sources. Secondary research also makes it possible to combine information from different sources to reach conclusions not suggestible by a single source (Stewart & Kamins, 1993, p. 2). Moreover, secondary data is less subject to biases that might occur if the researcher were to gather the information first-hand for a specific purpose (Boyd, Westfall & Stasch, 1994, p. 171).

Specifically, this research was intended to gather secondary data on the outcomes of e-health adoption for the various players across several countries that have made significant progress in e-health. Such data, published in both academic and trade journals, was compared and complemented with the primary data from qualitative research interviews so as to determine the extent to which the issues in e-health adoption as revealed by the conceptual model were prevalent in the cases studied, as well as to analyze and evaluate how these issues were dealt with in the cases. Such an approach, it was hoped, would likely be able to lead to answers to RQ2.

3.8.1.2 Qualitative Research Interviews

Interview is a mode of inquiry in qualitative research that can serve as one of the multiple sources of data for the case study method. The interview technique is one of the most common qualitative research methods used in healthcare research (Gill, Stewart, Treasure & Chadwick, 2008) and is probably the most suitable approach especially for investigating and understanding abstractions (Seidman, 2013). The e-health ecosystem, being nascent, can be thought of as an abstraction, something nonspecific and not concrete and is, therefore, liable to be influenced by perspectives, reflections and insights of healthcare consumers, policy-makers and industry players alike. Such attributes make it a suitable context for the interviewing technique.

Interviews may be of three types: structured, unstructured and semi-structured (Gill et al., 2008; Kvale & Brinkmann, 2009). Structured interviews have predetermined questions allowing for little or no variation during the interview process, and therefore offer limited scope to elicit elaborations. Unstructured interviews on the other hand do not have a fixed set of questions and often proceed from the initial response of the interviewee to a question. They are more like conversations and hence time-consuming. They may be relevant in contexts where very little information is available about the topic of interest while a great deal more is required. Semi-structured interviews, while providing a structure, also offer some flexibility to the researcher in respect of handling different respondents as the context dictates (Noor, 2008). Besides, they also allow divergence from the structure to pursue details (Gill, Stewart, Treasure & Chadwick, 2008). For the purposes of the current study which aimed to cover a list of topics as well as pursue emergent trajectories relevant to the topics concerned, semi-structured interviews seemed a reasonable choice as they balance structure and flexibility in right proportions.

3.8.2 Steps in Execution of Phase 2

The specific steps followed in executing phase 2 of this research are detailed below.

3.8.2.1 Step 1: Search for Potential Cases

Phase 2 began with a high level review of notable e-health developments around the world as recognized through literature reviews. A review of ten potential cases, well-known for their advanced healthcare systems, and recognized as pioneers in 'the use of technology for healthcare delivery', was undertaken.

3.8.2.2 Step 2: Defining the Case or Unit of Analysis

Based on the review in step 1, the unit of analysis for this research was defined as a 'nation-wide e-health ecosystem'.

3.8.2.3 Step 3: Design of Data Collection Instrument

A semi-structured interview was developed based on the game theoretical framework discussed in section 3.2, 'Major Barriers to E-health: A Game Theoretic View' which contributed to the conceptual model that was evolved in phase 1 of this research. The qualitative research interview was designed to be semi-structured with six interview questions. The interview template is given in Appendix C.

3.8.2.4 Step 4: Selection of Cases for the Study

During the initial review in step 1, it became increasingly apparent that secondary data was not easily accessible for some of the cases considered, while credible primary data was difficult to obtain for many. The nascence of e-health as well as the fact that several major initiatives are still ongoing could be attributed to the above-mentioned situation. In addition, being restricted to data sources in the English language also proved to be a constraint. By employing the purposive sampling strategies of **maximum variation** and **information richness** explained in section 3.4, 'Selection of Cases for Phase 2', the cases that were eventually selected for an in-depth study were Singapore's NEHR and the US' HITECH Program, both national-level e-health initiatives. One of the strategies behind such a selection, namely **information richness** was supported by the fact that primary data for the above-mentioned cases was accessible in the form of interviews with representatives of these e-health initiatives in addition to the availability of a variety of secondary data sources. Table 3.4 lists the profiles of the representatives

Table 3.4. Phase 2: List of target informants

Informant	Position/Profile	Remarks
#1	Managing Editor (and Healthcare Editor), FutureGov Asia, **Hong Kong**	#1 was kind enough to provide some leads for the case study.
#2	Advisor – IT Management, Pantai Hospitals, **Malaysia**	An email request to #2 to participate in our case study bounced back. Subsequently came to know that #2 had moved from Pantai Hospitals, Malaysia.
#3	CEO, HealthHiway, **India**	Sent a general email enquiry requesting for #3's direct email id, but did not get a response.
#4	Senior Vice-President and Chief Financial Officer – UPMC, **USA**	No reply.
#5	Technical Leader, TClouds (health cloud), IBM **Zurich** Research Lab, **Switzerland**	No reply.
#6	Director, Institute for Health Policy, Massachusetts General Hospital–Partners Healthcare System and Harvard Medical School – both in Boston**, USA**. He has been named National Coordinator for Health Information Technology.	Sent a general email enquiry requesting for #6's direct email, and received a reply from #6's Senior Executive Assistant, offering assistance on his behalf. However when participation in the case study was requested, it was turned down.
#7	Director of Medical Services, Ministry of Health, **Singapore**	No reply.
#8	E-Government Consultant at Prime Ministry, E-Health Coordinator at Ministry of Health, **Turkey**	No reply.
#9	Senior Researcher and Consultant at CEFRIEL, Milan Area, **Italy**.	No reply.
#10	Director of Information Technology, California Healthcare Foundation, **USA**	No reply.

(Contd.)

Table 3.4. (*Contd.*)

Informant	Position/Profile	Remarks
#11	Principal, Clinical Transformation Services, Information Systems Division, Ministry of Health Holdings, **Singapore**	No reply.
#12	IT Executive Director, UPMC, Pittsburgh, **USA**	#12's initial response was positive but subsequently the interest waned due to increased work commitments.
#13	Mayo Clinic, Rochester, **USA** Co-chair, SHARP 2.0	#13 responded to the email request but could not commit to an interview.
#14	Deputy Director, Health Care Division Ministry of Health and Social Affairs, **Sweden**	No reply.
#15	Professor of Health Informatics & Paediatrics, International Medical University, **Malaysia** Vice-President, Malaysian Health Informatics Association	No reply.
#16	Head of E-Government Group, Ministry of Health, **Bahrain**	No reply.
#17	**General Manager (Sales), Asia Pacific & Japan Napier Healthcare, Singapore**	**Conducted a face-to-face interview on 20/04/2015. Verified transcription of the interview on 26/06/2015**
#18	**ST Electronics (Info Software) Systems Pte. Ltd., Singapore**	**Conducted a face-to-face interview on 15/05/2015. Verified transcription of the interview on 05/06/2015**
#19	**Quintessence Business Solutions & Services, USA**	**Conducted an email interview. Response received on 23/08/2015**
#20	**Sutter Health, Sacramento, California, USA**	**Conducted a telephonic interview on 02/09/2015 and verified transcription on 03/09/2015**

Review of Key Concepts and Methodology

who were contacted, and tracks how they responded to the invitation for their participation in this case study research.

As revealed by this audit trail, the final response rate stood at a meagre 20% with only four of the twenty representatives eventually participating in the study. Two of these informants were professionally engaged with Singapore's NEHR (National Electronic Health Records) while the other two were engaged with the United States' HITECH (Health Information Technology for Economic and Clinical Health) program. Although the cases of these two countries are diverse in many respects, they share a strong history of early technology adoption (Stephanie, Tan, Morales-Arroyo & Sharma, 2011). It was therefore hoped that probing into the emerging e-health ecosystems of these countries would offer useful insights into the broad range of issues typically encountered in e-health initiatives.

It has to be recorded at this point that in spite of the authors' efforts over a substantial period of time to involve as many representatives from as many different cases as possible, the outcome was much less than what was desired. This may be partly due to the fact that e-health is an evolving field that involves interactions among several organizations from varying walks of the healthcare industry, and that the experts from one walk of the healthcare industry are only familiar with the changes taking place in their domain areas, and may not necessarily be conversant with developments in the other walks of the healthcare industry. Another likely reason is the intense nature of the interview envisaged which required, among other things, that the respondents assimilate the context and background information before responding to the questions posed. Thus, the initial, tentative, design had to be adapted to the emerging dynamics of the study with the result that only two case studies could eventually be done.

3.8.2.5 Step 5: Case Studies, Within-Case Analyses and Reports

The case studies utilized multiple sources of data such as literature reviews, documentation reviews, online searches and qualitative research interviews. Between the two cases selected in step 4, the study of Singapore's NEHR was designated the pilot case study in view of the ease and convenience of access it afforded to the informants.

The interview protocol that was followed to collect primary data is detailed below:

- A face-to-face meeting was scheduled with the Singapore-based informants for conducting the interview, and an email or a telephonic interview was conducted with the overseas informants based on their personal preference in terms of the mode (email vs telephonic) as well as the timing of the interview (for telephonic interviews). Follow-ups were made through emails or phone calls or a combination of both, for the purpose of clarification and elaboration.
- Interviewees were informed in advance that they might need to commit a little over an hour for the interview at their preferred timings and locations.

- The purpose of the study was explained to the interviewees in detail, and they were assured of the opportunity to review and suggest changes to the transcript of the interview, if they so desired. Only after their vetting was the interview included in the data, and analyzed.
- During face-to-face or phone interviews, research memos were written to flesh out concepts and patterns as they emerged.
- The face-to-face and phone interviews were transcribed and emailed to those informants who preferred to take a look at them before they confirmed the proceedings of the interview.
- If some informants wished to clarify or change their responses, a meeting or a phone conversation was set up with them for the purpose, and their interview was updated. Alternatively, where necessary, clarifications via email were sought as well as provided.
- The updated interview was again emailed to the informants for their verification and confirmation.
- Upon final confirmation from the informant, the interview was included in the data for the case study.
- The data thus collected was thematically coded. The coding scheme is presented in Appendix D.

The data collected from the various secondary research sources and the interviews, was thematically analyzed using the conceptual framework established in phase 1 of this research. The secondary research tapped into a wide range of documents – literature, news websites, industry journals, industry reports, blogs and parliamentary proceedings - to serve as sources of evidence. The data thus obtained was also compared with the data from the semi-structured qualitative research interviews for corroboration and triangulation. The within case analyses and reports for Singapore's NEHR and the US' HITECH program are presented in Chapters 5 and 6 respectively.

3.9 Phase 3: Cross-Case Analysis

The third and final phase of this research comprised a cross-case analysis of the two cases studied in phase 2, namely Singapore's NEHR and the US' HITECH Program. The primary focus of this analysis was to develop findings that were consistent across the two cases. Such an analysis involved a comparison of key themes that emerged from the two cases studied, thus allowing patterns to be perceived across the cases. These patterns enhanced the robustness of the findings and in turn, led to analytic generalizations with regard to the critical success factors for e-health implementations in view of the potential issues confronting e-health adoption. The cross-case analysis is presented in Chapter 7, and the conclusions, in Chapter 8.

3.10 Chapter Summary and Recap

This chapter provided a justification of the methodological approach adopted by this research and presented the roadmap followed by this research as well. The three phases in which this research was carried out were described above in detail. Phase 1 involved conceptualization of an e-health ecosystem that would be patient-centric, as well as, recognize issues that might jeopardize actualization of such an ecosystem. The sources of data for phase 1 were thematic literature reviews, industry observations and expert interviews. Phase 2 focused on testing the conceptual model evolved in phase 1, using case studies of Singapore's NEHR and the US' HITECH program. The two cases were selected through purposive sampling based on the strategies of *maximum variation* and *information richness*. Phase 2 relied on a variety of secondary data sources and qualitative research interviews for evidence. The data accumulated was thematically analyzed and within-case reports were prepared individually for the two cases. Phase 3 was concerned with a comparison of the key themes that emerged from the two cases studied in phase 2. A cross-case comparison such as this, illuminated common patterns and facilitated analytical generalization which, in turn, led this research to answers for RQ2 - the critical success factors for patient-centric e-health.

The next chapter marks the beginning of phase 1 of this research, the focus of which is to evolve a conceptual model of a patient-centric e-health ecosystem in response to RQ1.

CHAPTER

4

Conceptual Model of E-Health Eco-System

This chapter addresses RQ1 by constructing a conceptual model that embodies a patient-centric e-health ecosystem. This is proposed to be accomplished by identifying the potential key players in the ecosystem and comprehending their roles as well as the values they can create and capture in a patient-centric e-health ecosystem. This chapter is also intended to ascertain if this research is progressing with validity.

With these aims in view, various sources of data were collected through thematic literature reviews, industry observations and expert interviews. It is hoped that these sources of data would be adequate not only to arrive at a comprehensive set of potential key players who significantly contribute to the e-health ecosystem, but also to reaffirm the objectives of this research for their relevance and significance. These are a few essential first steps based on which further insights can be gained in terms of the ways in which the key players can drive e-health, the potential digital data flows among them and the values they can create and capture in the ecosystem in the process. The chapter ends with a synopsis of the key findings.

4.1 Data, Analysis and Discussion

4.1.1 Sources of Data for Modelling

For the purpose of finding insights to address RQ1, data from various sources such as thematic literature reviews, industry observations and expert interviews, were gathered. These data sources have been described in section 3.7, 'Phase 1: Development of conceptual model of a patient-centric e-health ecosystem'.

4.1.2 Key Players in E-Health

As regards RQ1, to ascertain the key players in e-health related literature that has emerged over the past decade or two was reviewed, and current market trends noted.

Conceptual Model of E-Health Eco-System

It needs to be noted that several authors (Parente, 2000; Joslyn, 2001; Aggrawal & Travers, 2001; Wen & Tan, 2003; Broderick & Smaltz, 2003; Walker et al., 2005; Konrad & Peter, 2007; Busch, 2008) have identified Patients, Providers, Payers and Vendors as the key stakeholders (players) in an e-health ecosystem. However, since their work preceded significant developments that took place in e-health later and fell short of a futuristic view, some potential key players were either unaccounted for or relegated to the background. Given below is a brief account of the aforementioned scholars' works and the gaps therein which helped guide further purposive sampling.

Parente's (2000) work focused on healthcare e-commerce and different business models including 'connectivity' models which function akin to the modern day HIE (Health Information Exchange), albeit for profit. Joslyn (2001) underscored the significance of patient-centric healthcare e-commerce, predicting a trend of 'personal medical records', which has now evolved into Personal Health Records (PHRs). Aggrawal & Travers (2001) made a case for the potential of web-based e-commerce in laying the foundation for effective and efficient transactions among the key healthcare market players, thereby improving information flows and reducing costs. The focus of their work was limited to business-to-business (B2B) and business-to-consumer (B2C) healthcare e-commerce. Wen & Tan (2003) examined possible opportunities and challenges facing key players in e-health so as to meet the needs of increasingly sophisticated consumers. They particularly drew attention to the dearth of investment in interactive technologies and e-commerce ventures from healthcare providers which can facilitate active and meaningful engagement of healthcare consumers. Although these studies primarily focused on e-commerce in healthcare, they incidentally provided significant inputs that helped to identify some key players in the e-health ecosystem. As Broderick & Smaltz (2003) had pointed out, e-health is not another name for healthcare e-commerce.

Besides the four players – Patients, Providers, Payers and Vendors, a potential fifth player, namely the infomediary, is also discernible in the systems these scholars describe, although no explicit mention of such a player is found in their works. Konrad and Peter's (2007) discussion of the potential role of an 'Intermediary' in enabling business processes integration in healthcare comes very close to the concept of infomediary, though the two concepts are not the same. Konrad and Haas described the 'Intermediary' as a 'broker between the cooperating parties' (p.5) and suggested the possibility of several intermediaries of the kind existing within an e-health ecosystem. In other words, the scope they envisioned for the role of the intermediary was rather narrow in the sense that it was not conceptualized as a single entity functioning as a conduit connecting all the key players in the e-health ecosystem. While examples of such infomediaries are abound in other industries, they are relatively new in e-health, and they are predicted to rise to prominence when they come to be abetted by cloud technology. Zahedi and Song (2008) described a health infomediary as a neutral online entity that could offer a range of services – illness and wellness related information, advice, guidance, assessment and referrals. However what is evident is that the role of a health infomediary has been steadily evolving in tandem with the advancements in technology and developments in e-health.

Of particular interest in this context are established technology companies that made their foray into e-health as cloud-based infomediaries. Some of these prominent players that pioneered into the health cloud space are Microsoft and Google. With their already heavy investments in cloud technology, they had the vantage position to extend their offerings to medical records services (Shimrat, 2009). Both Microsoft Health Vault (www.healthvault.com) and Google Health (www.google.com/health) shared the common goal of creating integrated online environments where one can create and store one's personal records, get information, find doctors, make medical appointments, communicate online, manage medications, share information with providers, perform tailored searches and do a lot more. Both players offered free web-based services that were consumer-centric in the sense that they put users in control of what went into their records, and who had access to those records. However, Google retired Google Health with effect from January 1, 2012, the reason being the inability of the product to create the sort of broad impact it was expected to make. Microsoft on the other hand, entered into a joint venture with GE Healthcare in the same year to launch a new company called Caradigm with a view to delivering an open, interoperable technology platform for healthcare organizations which, in turn, would augment the vision of a patient-centric healthcare system.

In their roles as infomediaries, players like Microsoft have the potential to function beyond traditional HIEs that merely enable the movement of health-related data among their multiple stakeholders. Microsoft may be ahead of the curve in this mission powered by open platform products that are vendor-neutral, interoperable and cloud-based.

Figure 4.1 shows a screenshot of the consumer interface for Microsoft Health Vault.

It is thus both evident and logical that an infomediary is an essential player in the e-health ecosystem who can foster effective connectivity among the other players. With the five key players already identified, an in-depth review of literature was undertaken to spot any other potential player not observed in the e-health ecosystem earlier.

Such probing revealed that it was not until 2003 that Regulators began to be acknowledged as part of the e-health ecosystem, when Broderick & Smaltz (2003) and later Walker et al. (2005) recognized the role of the government as not just one of the paying sources, but as a regulator of the e-health network. Broderick & Smaltz (2003) paid due attention to the significance of business to government (B2G) relationship in e-health in compliance with governmental regulations. Walker et al. (2005) discussed the role of the government in promoting interoperability standards to achieve a fully-standardized HIE that could not only reform healthcare but also potentially yield billions of dollars in financial returns.

Conceptual Model of E-Health Eco-System 81

Figure 4.1. Consumer interface for Microsoft HealthVault
Source: https://www.healthvault.com/sg/en

Another scholar, Busch (2008), suggested that the healthcare market has players at two levels – primary and secondary. At the primary level of the healthcare continuum are market players who use health information to provide patient care directly or indirectly (by supporting direct providers of patient care). These primary market players include Patients, Providers, Vendors and Payers. Secondary market players, on the other hand, use health information in roles other than direct and indirect patient care activities like public health, patient autonomy, quality assurance, safety, public policy, certification and standards, privacy, security, confidentiality and integrity (Busch, 2008). Although the above definitions served as useful criteria for this research to distinguish the primary from the secondary level e-health market players, it was observed that the primary e-healthcare continuum could not be complete without the role of an Infomediary who renders movement of health data across disparate organizations possible, and thus creates tremendous value in the e-health ecosystem. Another area this study did not subscribe to was Busch's classification of activities such as patient autonomy (empowerment), privacy, security, confidentiality, integrity, certification and standards as secondary, when in fact, these activities contribute to major regulations and standards that underpin the information exchanges among the primary e-health market players.

It is imperative for an e-health system to comply with these standards and policy regulations inasmuch as it is evident that the role of the Regulator as a primary level e-health market player cannot be overlooked. A case in point is the US healthcare industry where the federal government has tackled privacy and security concerns over electronic health information by extending the privacy and security regulations of the Health Insurance Portability and Accountability Act (HIPAA) to e-health vendors as well as Infomediaries such as Microsoft, to curb their commercial exploitation of health information (Blumenthal, 2009). The government has also enlisted the help of other private organizations to enforce standards for health information exchange. The Certification Commission for Healthcare Information Technology (CCHIT) for example, is a non-profit organization that establishes standards for an exchange of health information among the e-health players (Goldstein, 2009). Another non-profit organization collaborating with the US government in achieving widespread adoption of interoperable EHRs is Health Level Seven (HL7), which is involved in the development of international standards for the exchange, integration, sharing and retrieval of electronic health information.

Some other standards and vocabularies are SNOMED (The Systematized Nomenclature of Medicine), LOINC (Logical Observation Identifiers Names and Codes), and NCPDP (National Council for Prescription Drug Programs) (Soti & Pandey, 2007).

All in all, this study identifies six key players at the primary level: Patients (Healthcare Consumers), Providers, Payers, Vendors, Infomediaries and Regulators. At the secondary level are several public and private organizations focusing on activities such as research, surveillance and litigation. Table 4.1 presents a concept matrix of the key players in e-health as identified through literature reviews and industry observations.

When this provisional model was presented to the experts for their feedback, generally there was consensus among them that the model adequately reflected reality. The experts were in agreement with the six key players identified, and were predominantly of the view that the model reflected reality or was close to reality in the least. However there were other significant inputs from these experts that this research carefully considered for incorporation into the final model. Some of them are:

- Renaming the player "Patients" to "Consumers" or "Recipients" as the term "Patients" is rather narrow semantically. Not all recipients of healthcare services are patients as healthcare services cater not only to those who are ill but also to those who want to perpetuate their wellness. Besides, it is not only patients who use or contribute health data, but their next-of-kin may also do so on their behalf.
  "Patient would be just one of the categories in another broader category of Recipients. So I think it would be appropriate if you can name the category differently for e.g. Recipients or Customers/Consumers".

Conceptual Model of E-Health Eco-System 83

Table 4.1. Key players in E-Health: Concept matrix based on literature review and industry observations

Data source	Key players in E-Health					
	Patients	Providers	Payers	Vendors	Infomediaries	Regulators
1. Articles						
Parente (2000)	✓	✓	✓	✓		
Joslyn (2001)	✓	✓	✓	✓		
Aggrawal & Travers (2001)	✓	✓	✓	✓		
Wen & Tan (2003)	✓	✓	✓	✓		
Broderick & Smaltz (2003)	✓	✓	✓	✓		✓
Walker et al. (2005)	✓	✓	✓	✓		✓
Konrad & Peter (2007)	✓	✓	✓	✓	✓	
Busch (2008)	✓	✓	✓	✓		
Zahedi & Song (2008)	✓	✓			✓	✓
Blumenthal (2009)	✓	✓		✓		✓
Goldstein (2009)	✓	✓				✓
2. Industry observations						
Google Health (- 2011)	✓	✓	✓	✓	✓	
Microsoft Health Vault	✓	✓		✓	✓	

"You may want to include Next-of-Kin (NOK) who accompanied their parents or friends to seek treatment. Especially, for those who are elderly, information (treatment plan) usually flows to NOK rather than the patients. Most of the time, it is the NOK who make the decision on behalf of the patients".
""Patients" is a narrow definition".

This input, having been found valid and well-corroborated by literature (Wickramasinghe et al., 2005; Porter, 2009), led to the amendment of the label "Patients" in the provisional model to "Patients/Consumers" in order to be more inclusive.

- Research institutions and universities need special mention as they affect patient care directly with their treatment options.

  "Also need to consider research institutions, medical education systems, universities, scientists – they seem to affect patient care directly with their treatment options".

  "Research institutes and universities need a special mention".

Since the above-mentioned players only use health information in roles other than direct and indirect patient care activities (Walker et al., 2005; Busch, 2008), it was decided that they could not be considered primary e-health players.

Other Observations

In the course of the first author's interactions with experts, some observations emerged and were recorded in memos. Some of them are mentioned below for what they are worth.

- It was noted that use of healthcare-related terminologies was not standard across different cross-sections of the healthcare industry. For instance, while the term 'Providers' is commonly used in healthcare literature to refer more or less exclusively to healthcare providers, it was suggested that the meaning of the term could be extended to include patients' families and sometimes their domestic helpers as well, as they also play an active role in a patient's healthcare.
- The suggestion that 'traditional' or alternative healthcare practitioners such as Traditional Chinese Medicine, Ayurvedic practitioners and acupuncturists may also be considered for inclusion into the group of 'Providers ', sounded a little controversial. This necessitated a revisiting of literature in order to clarify and refine the role of a 'provider' in the e-health ecosystem.

The confirmatory e-health ecosystem is depicted in Figure 4.2.

4.1.3   Key Player Roles

Table 4.2 lists the key players in the e-health ecosystem, whose roles have been synthesized using data from literature review, industry observations and expert interviews.

Now that the key players in e-health have been identified and their roles understood through the various sources of data, the next step involves gathering field insights into factors that are believed to make e-health successful.

4.1.4   Critical Success Factors for E-Health

It was considered important to obtain a preliminary understanding of what the experts deemed to be critical success factors for e-health. In other words,

Figure 4.2. The validated E-Health ecosystem

this research sought to identify those elements which characterize or go into a successful e-health business model. To this end, the experts were consulted for their views on whether outcomes such as cost-efficiency and improved services are among factors that would make e-health successful. A summary of the experts' views which were divergent is provided below:

- While there was significant agreement that cost-efficiency and improved services are among factors that would make e-health successful, there was also some disagreement on this point which could not be ignored.
- Some experts expressed concern over the investment burden imposed on healthcare providers for the purpose of facilitating e-health. They seemed to have apprehensions that such a burden might prevent them from achieving cost efficiency. This sounded reasonable as ensuring adequate returns on provider investment is essential for the success of an e-health business model. One way this can be achieved is by incentivizing providers through government subsidies.

"The question now is whether the healthcare system can recover the IT investments that need to be made, through highly efficient and improved services."

"Some industry players (insurance companies, for example) are pushing doctors' offices to comply with requirements that are tremendously burdensome on the doctors' offices."

86   *Capturing Value in Digital Health Eco-Systems*

Table 4.2. Key players in the E-Health ecosystem and their roles

Players	Roles
Patients/Healthcare Consumers	Recipients of wellness- or illness-related health services for personal or next-of-kin's consumption – (Parente, 2000; Joslyn, 2001; Aggrawal & Travers, 2001; Wen & Tan, 2003; Broderick & Smaltz, 2003; Busch, 2008).
Providers	State-authorized providers of health services - clinical settings and professional staff that design, implement, and/or execute healthcare initiatives as part of wellness or illness programs (Parente, 2000; Joslyn, 2001; Aggrawal & Travers, 2001; Wen & Tan, 2003; Broderick & Smaltz, 2003; Busch, 2008).
Third-Party Vendors	Diverse market players who play a supporting role to the Providers in their provision of healthcare. These players include health IT vendors, medical equipment vendors, pharmaceutical vendors, transportation services, laboratories, legal systems, and billing agents (Aggrawal & Travers, 2001; Wen & Tan, 2003; Broderick & Smaltz, 2003; Busch, 2008).
Payers	Any entity that processes the claims payment transactions of healthcare episodes on behalf of plan sponsors. A plan sponsor is an entity that funds a health program – private insurance plans, government-sponsored plans, and employer-sponsored plans (Wen & Tan, 2003; Broderick & Smaltz, 2003; Busch, 2008).
Infomediaries	Organizations that act as mediators or brokers to facilitate information exchange among its network participants, gather pertinent health information from various sources, and syndicates, aggregate and distribute it to foster patient-centric healthcare (Morales-Arroyo & Sharma, 2009; Mettler & Eurich, 2012).
Regulators	Public and private organizations that develop capabilities for standards-based, secure and confidential exchange of health information to improve the coordination of care among the e-health market players (Broderick & Smaltz, 2003; Walker et al., 2005; Blumenthal, 2009).

"Providers lose out even in "steady-state" (not just in the initial stage). Govt. should subsidise EHR investment costs."

These views generally corroborated the findings from the literature reviews given in Chapter 3.

• Other views that were relatively less common but of interest nevertheless were that e-health is a politically driven decision, and that it would work only so long as the aggregate benefit to the society is optimal; it need not result in cost efficiency for every key player; on the same note, cost efficiency does not necessarily translate to lower costs for patients as they may be willing to pay a premium if they perceive value in a proposition.

- As a rule, the experts were of the view that e-health should be capable of bringing about adequate benefits including cost-efficiency and improved services failing which it cannot be said to serve any purpose. Undoubtedly, e-health promises significant benefits for most of the players as well as for the society at large, but the fact remains that it places a burden on the healthcare providers who are tasked with investing in EHRs though there is no guarantee of returns on their investment. The experts were thus cognizant of the possibility of misaligned incentives from e-health and emphasized the importance of addressing the likely uncertainties and imbalances to ensure that all players get their fair deal. These findings, significant as they are, underscore the relevance and significance of the present study.

4.1.5 Value Analysis of the E-Health Ecosystem

This section focuses on the potential values that the six key e-health players identified previously can create and capture in a patient-centric e-health ecosystem. This part of the research activity involved three sequential steps. The first step aimed at developing a simplified model of the significant digital data flows in the e-health ecosystem. Based on the insights that emerged from the first step, the second step consisted in assessing and articulating the potential values generated in the e-health ecosystem using the ADVISOR business modelling framework. The third step was intended to identify and corroborate the values generated in the e-health ecosystem, using data sources such as thematic literature reviews and expert interviews.

4.1.5.1 E-Health Digital Data Flows: A Simplified Model.

A simplified model of the digital flows or exchanges in e-health is derived based on an understanding of the e-health ecosystem, its key players, their roles and business model arrangements. In this model, the players on the e-health network primarily assume two roles: one of 'producers' of health data which could be supplied to the network, and the other of 'consumers' of health data from the network. These two roles are not mutually exclusive and may often co-occur. For instance, the information produced and supplied by a physician in the form of EHRs may be consumed by another healthcare provider in the network, say, a specialist or a hospital, who then makes diagnoses with the health data, in turn producing more data, which could be supplied to other players such as insurers. Likewise, the information produced and supplied by a patient or a healthcare consumer in the form of PHRs may be consumed and edited by others in the network authorized to do so. Such exchanges are facilitated by an 'infomediary', yet another role that enables movement of health data among the various players in the network through syndication, aggregation, and distribution of health information in their central repository, thus providing added value to the ecosystem.

In essence, an e-health infomediary brings together or bridges various players who would otherwise belong to a highly fragmented market, and facilitates seamless digital data flows among them. Figure 4.3 shows some of the significant digital data flows that are enabled by the infomediary in the e-health network.

E-health digital data flows fall into two categories: information-based and transaction-based flows. 'Information flows' denotes the information supplied to and consumed by the key players in the e-health network. 'Transaction flows', on the other hand, means business transactions among the key players. For instance, information flows may include exchange of EHRs between providers, or between a provider and a payer or vendor. Similarly, PHRs can also be exchanged by a patient or consumer with providers, payers and vendors. Transaction flows involve e-commerce and may include instances like a provider purchasing medical supplies from a vendor, or a patient or consumer purchasing products from a vendor or services from a provider.

In addition to mapping the digital flows, the added values that accrue to every player on account of these digital exchanges (enabled by the e-health platform), have also been identified from the thematic literature. Figure 4.3 sums up these

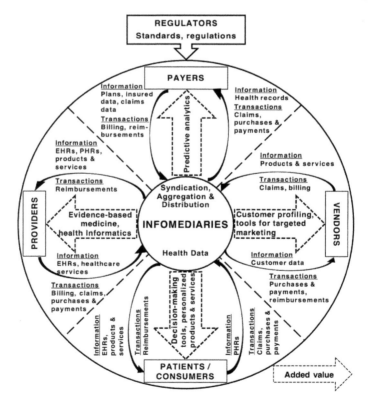

Figure 4.3. E-health digital flows: A simplified model

values. The modelling of digital data flows in the e-health ecosystem is based on an understanding of the key player roles and business model arrangements with the resultant model reflecting the significant data flows that are possible. Determining these digital data flows is crucial inasmuch as it sets the stage for identifying the values that can potentially be generated in the e-health ecosystem.

4.1.5.2   Value Analysis using ADVISOR Business Model

The conceptual framework exploited to analyze the e-health ecosystem for the values created and captured is the ADVISOR Business Model. This model is an extension of the VISOR Business Model proposed by the Centre for Technology & Management of the Marshall School of Business at the University of Southern California (El Sawy, Pereira & Fife, 2008). The extended ADVISOR framework was developed by the Special Interest Group on Interactive Digital Enterprise (SIGIDE), Nanyang Technological University (Sharma, Pereira, Ramasubbu, Tan & Tschang, 2010). The VISOR framework helps comprehend how players in the digital business industry may evaluate and capitalize on the emergence of a new technology or service offering. It comprises five variables that need to be considered by players: **V**alue Proposition, **I**nterface, **S**ervice Platform, **O**rganizing Model and **R**evenue/Cost Sharing. The ADVISOR framework adds two variables to VISOR namely **Adoption by Consumers** and **Disruptive Innovation,** and suggests that the value created and captured by a player in the digital marketplace is a function of the following seven key design parameters:

Value Proposition for the Customer	Compelling value of the digital product or service provided by a player.
Interface or the "Wow" Interface Experience	Easy to use, simple and convenient user interface for the successful delivery of the product or service.
Service Platforms to Enable Delivery	Platform(s) that support the business processes and relationships needed to deliver the product or service.
Organizing Model for Processes and Relationships	Value chains, business processes and partner relationships to ensure the effective and efficient delivery of the product or service.
Revenue / Cost Sharing for Partners	Business justification for the investments in providing the product or service and a fair return of revenues for all players involved.
Adoption by Customers	*How customers in effect co-create additional value by using or consuming the product or service*
Disruptive Innovation	*Impact of new technologies and business arrangements on the market.*

It is posited that the total value of an e-health ecosystem is a function of the above components.

It may be helpful to explain why the ADVISOR business modelling framework is chosen from among the many that are available such as the PARTS framework, Porter's 5-Forces, the BCG Box, and the Diamond Model. A business model design is a significant decision not only for new firms, but also for existing firms that need to rethink their current models to make themselves relevant for the future (Zott & Amit, 2010). This is true especially of the digital business industry characterized by transformational forces that lead to turbulences and vulnerabilities in the market (Sharma et al., 2010), very often resulting in what Christensen (1997) termed 'disruptive innovation'. Disruptive innovation is a process that has positively transformed the status quo for customers in several industries by effectuating convenience and affordability. However one of the industries that remain largely untouched by disruptive innovation is the healthcare industry (Hwang & Christensen, 2008). In spite of the numerous technologies being introduced into the industry on a regular basis, healthcare is still no more accessible or affordable than in the past. As rightly observed by Hwang and Christensen (2008), it is time to ask how can we make healthcare more affordable rather than how we can afford healthcare (p. 1329). They further note that only a business model that is designed to offer increased convenience and affordability by taking advantage of disruptive innovations can create tremendous value in healthcare.

A business modelling tool that explicitly embeds in its design the notion of 'disruptive innovation' as a key parameter is the aforementioned ADVISOR. No wonder it is considered an apt tool to analyze and articulate how the various players in the e-health industry may deliver value to their co-players through both cooperative and competitive means (co-opetition).

Table 4.3 presents an analysis of the e-health ecosystem using the ADVISOR business model framework. The analysis is from the perspective of each of the key players identified in the e-health ecosystem with the exception of the Regulators, who are not-for-profit entities and who therefore cannot be subject to a business model framework. The conceptual framework is used to comprehend how the key players in the e-health ecosystem may respond to e-health, evaluate the potential of e-health in terms of the new business opportunities it presents, and strategize to capitalize on these opportunities. This analysis is backed by a comprehensive review and understanding of the various existing e-health business models presented in section 3.2.6, 'Business Models in E-Health' and the roles of the key players that have been identified are presented in section 4.1.3, 'Key Player Roles'. The framework necessitates critical thinking along the seven design parameters to identify the key issues involving e-health from every player's perspective and determining how best the player would address these issues. It is hoped that such an assessment of the ecosystem for its value potential can lead to evolving an efficient and fair (win-win) arrangement among the players so that each of them derives returns (value captured) that are consistent with their added values (value created). After all, creating and capturing values are fundamental functions a player must perform to remain in the game (Shafer et al., 2005).

Table 4.3. Analysis of the E-health ecosystem using ADVISOR

	Key players				
	Patients (Consumers)	**Providers**	**Payers**	**Vendors**	**Infomediaries**
Value Proposition	Patient-centric, affordable healthcare facilitated by a longitudinal health record and other healthcare decision-making tools	Evidence-based medicine facilitated by EHRs	Efficient claims handling and reimbursements and better risk modelling	Direct, targeted marketing of products and services to healthcare customers	A "computing as a utility" platform facilitating exchange and reuse of health information which reduces costs and improves efficiencies for all
Interface	Web interface that allows for interoperability and exchange of data residing with disparate organizations				
Service platform	Cloud computing technologies with high security features like encryption, authentication and access levels to protect health data				
Organizing model	Standards and certifications for interoperability of health information systems and seamless exchange of health data subject to appropriate access levels (e.g. HL7, Dicom etc.) Regulations to protect privacy and security of health data and curb its commercial exploitation (e.g. HIPAA)				
Revenues/ Cost	Cost: Free or small subscription fee Revenues: Usually none, with some possibility for C2C commerce	Cost: Investment in EHR systems, training costs, maintenance/ upgrading costs Revenues: B2B and B2C e-commerce, reimbursements	Cost: Investment in payer information system, training costs, maintenance & upgrading costs Revenues: B2B, and B2C e-commerce	Cost: Investment in information systems, training costs, maintenance & upgrading costs Revenues: B2B, B2C e-commerce, reimbursements	Cost: Investment in cloud infrastructure for standards-based, secure exchange of health data. Revenues: Subscription fees, advertisement revenues, e-commerce transactions

(Contd.)

Table 4.3. (*Contd.*)

	Key players				
	Patients (Consumers)	**Providers**	**Payers**	**Vendors**	**Infomediaries**
Adoption by users	Co-creation of value through PHRs, a source of empowerment in healthcare decision-making	Co-creation of value through EHRs which improve efficiencies and reduce medical errors and resulting litigation costs	Co-creation of value through aggregation of insurance and claims data which facilitates superior risk modelling	Co-creation of value through targeted marketing which significantly reduces direct marketing costs	Co-creation of value through syndication, aggregation and distribution of health data which creates new markets and business opportunities
Disruptions	Single electronic interface for the entire continuum of healthcare management in lieu of multiple resource-consuming face-to-face interfaces	Electronic interface - with patients in lieu of resource-consuming face-to-face interface - with payers and vendors for efficient and transparent transactions	Electronic interface for insurance claims and purchase of plans in lieu of resource-intensive paper-based transactions	Single electronic interface to access and target the gamut of healthcare customers - healthcare providers, patients, insurers	A well-networked electronic healthcare marketplace on cloud in lieu of the conventional, fragmented marketplace

Conceptual Model of E-Health Eco-System 93

From the foregoing analysis of the e-health ecosystem using the ADVISOR framework (Table 4.3), it is apparent that, if organized by the right business model, e-health holds immense potential for the future. The businesses of the key players are interdependent insofar as every player potentially creates an 'added value' that benefits the other players, thereby giving rise to a compelling proposition in the health ecosystem to collaborate to create and capture 'value' rather than compete for 'dollars'.

4.1.5.3   A Qualitative Analysis of Values in the E-Health Ecosystem.

Supported by such insights from ADVISOR modelling, data collection was directed towards identifying the significant values that could be created and captured by players in the e-health ecosystem conceived in this research. This was accomplished through extensive literature reviews the findings from which are compared and corroborated with inputs from expert interviews.

Data Source 1: Thematic Literature Reviews

Patients (Consumers)

E-health holds great promise for empowerment of patients who are "the largest and most important stakeholder group" (Hill & Powell, 2009). It serves to open doors to competitive markets, which is bound to increase healthcare choices and lower healthcare costs. Equipped with ubiquitous access to PHRs, EHRs and tools for informed healthcare decision-making and guided health promotion behaviors (Burkhard, 2009), healthcare consumers will have complete freedom to choose and switch among multiple care providers while seeking to satisfy their unique needs for quality, service, and price (Joslyn, 2001).

A 2007 survey of US consumers by Accenture found that healthcare consumers regard the presence of EHRs as an important factor in their choice of a physician ("Provider-Led Health Insurers - Healthcare Consumer Satisfaction - Summary - Accenture", 2007). Another study by User Centric (Peters et al., 2009) on the comparative usability of two cloud-based PHR applications (Google Health versus Microsoft Health Vault) reported that the majority of its study participants found PHRs useful and expressed an interest in creating and maintaining their own PHRs. A more recent study by Zieth et al. (2014) found that patients expressing dissatisfaction with their current healthcare providers were more likely to prefer an EHR-integrated PHR that would not only give them access to their health information, but also empower them to take control of their healthcare related decisions. Such trends reflect the rise of consumerism which, when coupled with policy imperatives by governments, forces healthcare providers to focus on the needs and demands of consumers, and consequently make hefty investments in EHR systems.

Providers

EHR investments may seem a burden on the healthcare providers in the short term, but the value the providers as well as the other players can leverage from this investment over the long term appears compelling. It has already been discussed how cloud technologies may be explored by healthcare providers to ease their EHR investment burden. When e-health fosters evidence-based medicine (Busch, 2008) with the support of medical informatics, it enhances the quality of care, improves operational efficiencies; reduces medical errors (Miller, West, Brown, Sim & Ganchoff, 2005; Chaudhry et al., 2006; Hill & Powell, 2009); facilitates training and education of physicians and cost-effective e-procurement of medical supplies (Wickramasinghe et al., 2005); increases reimbursements (Berthold, 2008) and provides access to a larger pool of healthcare consumers by opening up new business opportunities (e.g. authoring personalized healthcare plans).

Payers

The e-health benefits that accrue to Payers are also significant - improved efficiency in claims processing resulting in cost savings (Wen & Tan, 2003; Walker et al., 2005; Menachemi & Collum, 2011), easier implementation of regulations such as HIPAA (Wen & Tan, 2003), reduced wasteful reimbursements on account of redundant tests (Goldstein, 2009), online sale of health insurance bypassing agents (Whitten, Steinfield & Hellmich, 2001) and research and analytics enabled by the very large stores of data aggregated in the e-health network (Wen & Tan, 2003).

Vendors

The diverse range of healthcare Vendors also adds to the list of benefits from e-health in terms of the enormous potential for B2B and B2C e-commerce opportunities it offers (Payton, 2003). They enjoy better visibility of their offerings, and gain direct access to their customers which in turn decreases their marketing costs (Wen & Tan, 2003). Additionally, they are also equipped with tools to do targeted marketing of their products and services to prospective customers.

Infomediaries

For the infomediary, the value in e-health is the opportunity to syndicate, aggregate and distribute the massive amounts of health data in the network referred to as 'big data' (Kayyali, Knott & Van Kuiken, 2013; DeNardis, 2014), in order to cater to the varying needs of the other players, thereby creating new business opportunities and markets (Morales-Arroyo & Sharma, 2009; Mettler & Eurich, 2012). By facilitating such exchange and reuse of health data (Brailer, 2005), it improves efficiencies and reduces costs for every player in the network while simultaneously generating revenues for itself through such value creation. Infomediaries already in the cloud (e.g. Microsoft Health Vault) may well emerge the most dominant infomediaries in the e-health ecosystem, with the take-off of cloud computing technologies in the healthcare sector.

Regulators

The regulators, although non-profit entities, can also create and capture value in the e-health ecosystem even if it may not be in the form of direct economic returns. They view e-health as a means to improve the quality of healthcare

Conceptual Model of E-Health Eco-System **95**

(Blumenthal, 2009). To achieve this objective, they drive the standards required to facilitate interoperability among health systems so that health data can be exchanged and reused (Goldstein, 2009), and stipulate regulations to protect the privacy and security of health data, to prevent any unauthorized exploitation of such data by other players in the network (Blumenthal, 2009). Thus, the regulators are in a position to improve the health of populations, augment the efficiency of healthcare systems, and lower healthcare expenditure (Blumenthal, 2009). In addition, regulators may also offer financial incentives to healthcare providers to encourage adoption of EHR systems by reducing their financial burden in such investments (Jones, 2012; Jacob, 2013).

Data Source 2: Expert Interviews

There was a broad consensus among the experts when they were asked to confirm whether the sources of values in the e-health ecosystem had been adequately identified and defined. Some additional values suggested by a few experts for consideration in this research, were:

- Time effectiveness
- Capability to prioritize attention to patients
- Provider education

These suggestions led to further data collection for corroboration, and were eventually included into the list of significant values in the e-health ecosystem presented in Table 4.4.

Yet another insightful suggestion was to explicitly map out the contributors (creators) of values and their corresponding beneficiaries (capturers). After considerable thought, it was decided that such a mapping might be too complex to prove useful. This is due to the possibility that some of the values may be co-created by multiple players and some of the values may have multiple beneficiaries.

Based on the assessment of the key players' potential for value creation and value capture in the e-health ecosystem using the ADVISOR framework; the literature-backed identification of the values generated in the ecosystem; and the inputs elicited through the expert interviews, the significant values created and captured by every key player in the ecosystem are juxtaposed in Table 4.4. It is to be noted that these values may be tangible or intangible, and are not exhaustive.

Analysis and Discussion

Apparently, there are significant values to be captured for every key player in the e-health ecosystem thus encouraging their participation in the ecosystem. The ecosystem seems to present opportunities for every key player to create values that benefit the other players, and correspondingly capture sufficient values to justify their participation and value creation in the ecosystem. As noted by Michel (2014), it does not suffice if a business keeps innovating itself to create new values for its customer; it should also revisit how it captures values so as to be adequately incentivized for its value creation. Thus, it is not unreasonable to surmise that a fair, efficient, stable and sustainable e-health ecosystem would

Table 4.4. Validated values created vs. values captured

Key players	Values created	Values captured
1. Patients/ Consumers	□ Health data (Neupert, 2009) □ Personal health records (Peters et al., 2009) □ Healthcare e-commerce (B2C, C2C) (Parente, 2000; Eysenbach, 2001) □ Health experience sharing in peer-to-peer communication (Wickramasinghe et al., 2005; Calvillo et al., 2013)	□ Ubiquitous access to health records (Burkhard, 2009) □ Patient-centric healthcare (Burkhard, 2009) □ Increased healthcare choices (Joslyn, 2001; Powell & Laufer, 2010) □ Lower healthcare costs (Wen & Tan, 2003; Tang et al., 2006) □ Timely intervention (Archer et al., 2011) □ Empowerment to make informed choices (Purcarea, 2009; Calvillo et al., 2013)
2. Providers	□ Electronic medical / health records (Hill et al., 2007) □ Improved quality of care (Walker et al., 2005) □ Tele-healthcare delivery (Hill & Powell, 2009) □ Evidence-based medicine (Busch, 2008) □ Preventive healthcare (Chang et al., 2009) □ Medical informatics (Raghupathi & Kesh, 2009) □ Reduced medical errors (Vishwanath & Scamurra, 2007)	□ Improved operational efficiencies (Wen & Tan, 2003; Ahern, 2007) □ Enhanced cost-effectiveness (Vishwanath & Scamurra, 2007) □ Time-effectiveness enabled by automatic capture of patient data (McKesson Corporation, 2007) □ Capability to prioritize patients based on the severity of their condition e.g. triage services (Konrad & Peter, 2007) □ Training and education of physicians (Wickramasinghe et al., 2005) □ Cost-effective e-procurement of medical supplies (Wen & Tan, 2003) □ New business opportunities created by the eHealth network (B2B, B2C) (Parente, 2000) □ Reduced risk of litigation (Couch, 2013)
3. Third-Party Vendors	□ Direct access to products / services for customers (Wen & Tan, 2003; Atkinson et al., 2009) □ Personalized products and services facilitated by data aggregation (Parente, 2000)	□ Electronic interface to access and target the gamut of healthcare customers – healthcare providers, patients, insurers (Wen & Tan, 2003) □ Decreased marketing costs (Wen & Tan, 2003) □ Reduced R&D costs (Wen & Tan, 2003)

	□ R&D data (Kayyali et al., 2013)	□ Increased business opportunities (such as B2B, B2C) due to direct access to a large pool of healthcare customers (Payton, 2003; Wickramasinghe et al., 2005)
4. Payers	□ Electronic interface for claims management (Walker et al., 2005) □ Electronic interface for direct purchase of plans by customers bypassing agents (Whitten et al., 2001) □ Research and analytics facilitated by data aggregation (Wen & Tan, 2003) □ Targeted information delivery to customers facilitated by data aggregation (Parente, 2000)	□ Improved efficiency in claims handling (Wen & Tan, 2003; Walker et al., 2005; Menachemi & Collum, 2011) □ Cost savings (Walker et al., 2005) □ Easier implementation of healthcare regulations (Wen & Tan, 2003) □ Increased business opportunities (i.e B2B, B2C.) due to direct access to a large pool of healthcare customers (Parente, 2000)
5. Info-mediaries	□ Total digital health systems (Raghupathi & Kesh, 2009) □ A digital platform to facilitate exchange and reuse of health data among players in the healthcare market (Morales-Arroyo & Sharma, 2009; Mettler & Eurich, 2012) □ Improved efficiencies for players in the healthcare market (Wen & Tan, 2003) □ Reduced costs for players in the healthcare market (Wen & Tan, 2003) □ Big data (Kayyali et al., 2013; DeNardis, 2014)	□ New markets brought about by syndication, aggregation and distribution of health data (Morales-Arroyo & Sharma, 2009) □ New business opportunities and revenues to be tapped in the form of subscription fees, advertisement revenues, e-commerce transactions (Parente, 2000)

(Contd.)

Table 4.4. (*Contd.*)

Key Players	Values created	Values captured
6. Regulators	☐ Incentives to facilitate EHR adoption and health information exchange (Jones, 2012; Jacob, 2013) ☐ Regulations to protect privacy / security of health data (Blumenthal, 2009) ☐ Standards for interoperable healthcare systems to facilitate exchange and reuse of health data (Goldstein, 2009)	☐ Improved health of populations (Blumenthal, 2009) ☐ Improved efficiency of healthcare systems (Blumenthal, 2009) ☐ Lower healthcare expenditure (Hill & Powell, 2009)

Conceptual Model of E-Health Eco-System 99

be one that justifies the participation of every player by ensuring that there are compelling values they can capture from the network in exchange for the "added values" they create. However, the actual outcomes of such a proposition may vary from player to player, thus posing the question of whether the values captured (values added) justify the values created (added values) for all the players involved. After all, value depends on which stakeholder's perspective it is viewed from (Rudin et al., 2014). In recognition of this fact, expert views were sought to determine if there is any perceived imbalance in values-created versus values-captured for any player(s) in the e-health ecosystem, and to understand whether fairness and efficiency were considered to be key design parameters for a sustainable ecosystem. Any new issues brought up by the experts were compared with literature for conformity or contrast.

While some experts were of the opinion that the values created and captured by the key players are congruous, the others held divergent views as shown below:

Players perceived to create more value than they capture:

- Providers were widely considered to be creating more value than they capture from the ecosystem because the onus of the heavy financial burden in terms of EHR investments is directly on them. E-health, in the view of the experts, has yet to evolve, firstly to the point of preserving the productivity of the Providers, and then to the point where the Providers can actually start reaping the benefits from their heavy investments.

   "Providers – from adopting an e-health mode of treatment delivery they are yet to reap the full benefits of the high investments. E-health at this point in time is yet to evolve to preserve the productivity of providers".

   "The value created for the patients will be bigger than value captured as they will then be able to access different providers of health care".

   "Providers create more value than they capture because of EHR investments".

These opinions are consistent with the findings from our review of the literature in Chapter 2 and conceptual themes in Chapter 3.

- Regulators were considered by some experts to be the players that creates the maximum value in the e-health ecosystem. Such significant contributions of the Regulators is expected to continue until the ecosystem stabilizes and healthcare related indices reflect better measures as a direct result of adoption of e-health.

   "Regulators/Government create maximum value until indices begin to show better health values because of the adoption of e-health".

   "The government will create the greatest value as it will lower the overall cost of the public health care".

That the regulators play a significant role in creating values in e-health is indeed true as they are the entities tasked with improving outcomes in healthcare, a public good, through appropriate regulations, standards and other actions (Accenture, 2012).

Incidentally, some experts were of the view that the Regulators might capture the least value which however may not be true. By creating values in healthcare enabled by e-health, regulators are not only able to achieve population health goals (Diamond, Mostashari & Shirky, 2009), but are also able to retain their legitimacy as regulators since the value they create is a yardstick for the public to assess their performance (Kelly & Muers, 2002).

- Some dissenting viewpoints on the role of an infomediary may also be mentioned here. Some experts held the view that Health IT Vendors create the greatest value because they act as the conduit through which products and services reach Patients in the most effective and efficient manner.
  "We (third-party vendors) are the conduit through which the services reach the patients in the most effective and efficient manner".
  It is inferred that the 'IT vendors' referred to in this context are, as a matter of fact, the 'infomediaries' as described in this research. One of the key propositions of such an infomediary is to meet the market demand for value-added services that would enable its customers 'to find what they are looking for', especially in the face of the increasingly ubiquitous Internet (Konrad & Peter, 2007, p. 105).

In contrast, some experts were of the opinion that the infomediary creates the least value.

The contrasting scenarios indicate that the role of an infomediary is not clearly understood. It is likely that the role of the infomediary may have been construed as similar to that of a typical IT vendor whose role is limited to the institutional boundaries of its customers (e.g. Providers). As a matter of fact, the role of the infomediary spans multiple organizational and even industrial boundaries in its bid to forge meaningful connections among the various stakeholders in its network. Consequent to this ambiguity regarding the role of the Infomediary, the term was redefined so as to improve clarity.

Players perceived to capture more value than they create:

- By and large, patients (consumers) were considered to be the players capturing the greatest value from e-health. In fact, they were viewed as capturing more value than they create, and quite rightly so, because they are the intended beneficiaries of any health system (Hill & Powell, 2009).
- Another player that was thought to capture more value than it creates is the EHR vendor. This is because the EHR vendors are often known to capitalize on the system integration opportunities arising from the lack of interoperability across hospitals' IT systems.

The above-mentioned observation is alluded to by Kellermann and Jones (2013) who suggest that the lack of progress on interoperability may be a result of the resistance from not only the providers but also the Health IT vendors. This is because, enforcement of interoperability will lead to standardized health IT systems which will not require the intervention of an IT vendor to be able to exchange information among them. This, in turn, will result in loss of business opportunities for EHR vendors who are known to charge hefty 'interface fees' to facilitate interoperability across silo systems (Tahir, 2014).

- A rather unique but noteworthy perception, though not commonly encountered, is that Payers create the least value since the adoption of e-health and other technologies has not resulted in reduced plan premiums and improved benefits for consumers.

It is true that there is no strong demonstrable evidence of the benefits of e-health to date. This may have to do with the fact that the value of Health IT accrues over time (Rudin et al., 2014) and for any transformation to be achieved in healthcare, continuity and consistency are key (Britnell, 2015).

Perceptions of fairness and efficiency

The predominant view among the experts seemed to be that e-health will come to fruition only if there is fairness and efficiency in the ecosystem, and that an imbalance between values created and captured for any player will render e-health unsustainable in the long run.

A different view though rarely encountered is that e-health may still be sustainable if the outcome is Pareto-efficient, i.e., no player is worse off than status quo. This view is in conformity with the line this research has taken and is worth exploring in further research as well.

Figure 4.4 shows a concept map of the key findings related to the e-health ecosystem and the issues therein, some of which require further investigation. These key findings represent the key concepts that have so far emerged in this research.

4.2 Caveats

The modelling of the e-health ecosystem was primarily based on literature reviews, observations of market developments, and expert interviews. The key players identified are by no means exhaustive, but can be classified into six broad categories as discussed in this chapter. Such a classification is not merely for the purposes of ease or simplicity, but rather to keep this research focused on the need to investigate the critical issues facing the e-health ecosystem. Likewise, the data flows identified among the key players are non-exhaustive but reflect those that are significant.

4.3 Chapter Summary and Recap

The purpose of this chapter was to address RQ1 through the identification of the key players in a patient-centric e-health ecosystem, their roles and the potential values they can create and capture in a patient-centric e-health ecosystem through digital data flows. Also embedded within the scope of this chapter was the objective to validate the relevance and significance of this research. To help reach

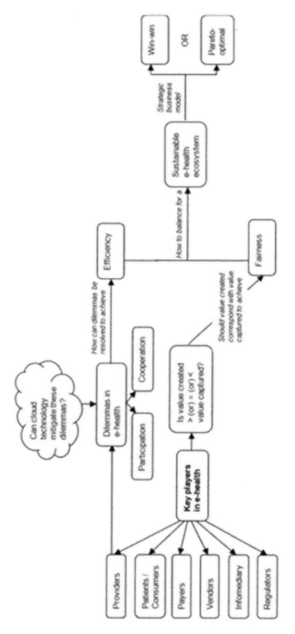

Figure 4.4. A concept map of findings about the e-health ecosystem

these targets, multiple sources of data such as thematic literature reviews, industry observations and expert interviews were deployed. It may not be an exaggeration to say that the findings based on these data sources supplemented by ADVISOR business modelling have been insightful. Supplemented by updated research, these findings have been instrumental in augmenting the provisional models developed early on in this research.

The findings so far have helped establish the fact that there are six key players in an e-health ecosystem without whom, the ecosystem may not be able to generate sufficient values for its balance and sustenance. Also, evident through the findings is the fact that the e-health ecosystem offers the potential for every player to capture values which may otherwise be inaccessible to them; however, it may only be sustainable if it is fair and efficient from every player's perspective. At the least, no player should be worse off than status quo because of their participation in e-health. In other words, the minimum requirement for an e-health system may be to be Pareto-efficient. Currently, there is a perceived imbalance in the ecosystem because of the financial strain it saddles the Providers with. The Providers are seen as having to contribute more value to the ecosystem than they can capture. This clearly tilts the balance in the ecosystem to the disadvantage of the Providers, and will lead to a less than Pareto-efficient ecosystem which may not be sustainable. Our research sets out to further investigate how a balanced and sustainable ecosystem may be achieved, and if the cloud technology has the potential to render it possible. With these steps accomplished, RQ1 may be said to have been addressed, and phase 1 of this research completed. Moreover, the relevance and importance of this research also stands validated through the triangulation of the various slices of data which imply that misaligned incentives among the key stakeholders of e-health may pose a challenge for its successful implementation. Besides, the discussion embodied in this chapter helped outline the first ever key steps needed to set the study into focus and look for guidelines for seeking answers to RQ2.

The next chapter marks the beginning of phase 2 of this research. It presents the case study of Singapore's NEHR, an ongoing e-health initiative. This case study forms the pilot study undertaken to test the conceptual model of the e-health ecosystem developed in phase 1 of this research.

CHAPTER

5

Within-Case Analysis of Singapore's NEHR

This chapter marks the beginning of phase 2 of this research which was concerned with validating the conceptual model of a patient-centric e-health ecosystem developed in phase 1. Such validation was intended to be accomplished through two descriptive case studies namely Singapore's NEHR and the US' HITECH program. The rationale behind the choice of these cases for study has been elaborated in section 3.4, 'Selection of Cases for Phase 2'. Of the two cases chosen, Singapore's NEHR was designated the pilot study owing to the relatively easier accessibility it afforded to the informants. It was hoped that this pilot study would cast light on improvement opportunities in the case study design. This chapter presents the within-case analysis and report of Singapore's NEHR, the pilot case study, and ends with a brief account of the lessons learnt in the course of this case study and key conclusions from the study.

5.1 Introduction

Singapore's healthcare system is a two-tier system which involves participation of both public and private sectors in the provisioning as well as the financing of healthcare services (Lim, 2005). The distribution of Singapore's healthcare services across the various sectors is depicted in Figure 5.1. The focal point of Singapore's healthcare financing system is the 3M scheme – Medisave, Medishield and Medifund. This is based on the philosophy of shared responsibility. This means the government would subsidize healthcare costs, but the people will still have to share the costs of the healthcare services they use so as to keep the healthcare system viable and sustainable. Ranked the world's 'healthiest country' in 2015 alongside Iceland and Sweden ("World Health Statistics 2016: Monitoring health for the SDGs", 2016) and having a healthcare system which was the world's 'most efficient' in 2014 (Chen, 2014), Singapore, facilitated by its current position at the forefront of technology, is well-poised to successfully face the challenge of transforming its healthcare further. It is one of the few countries that ranked high on e-health preparedness and hence was predicted to successfully implement e-health (Wickramasinghe, Fadlalla, Geisler & Schaffer, 2005). This is despite Singapore's total health expenditure as a percentage of its Gross Domestic Product (GDP) being 4.9 in 2014 ("World Health Statistics 2016: Monitoring health for the SDGs", 2016), relatively far lower than that of other advanced countries.

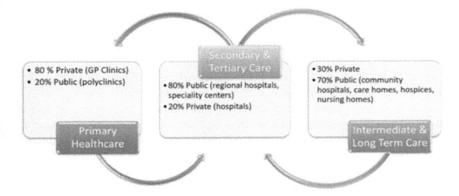

Figure 5.1. Singapore's distribution of healthcare services

5.2   Singapore's Healthcare Transformation Journey

Singapore has reached its current position as a world leader in healthcare only after a long and hard journey. A revisit to the history of healthcare in Singapore since 2000 may be in order in this context. To make the healthcare environment competitive, Singapore's public healthcare system was restructured into two major clusters, namely Singapore Health Services (SingHealth) and National Healthcare Group (NHG) in 2000 (Chang, 2010). However, this inter-cluster competition did not prove very helpful as it was perceived to be artificial, and existed only in form but not in substance (Lim, 2005). Such an arrangement also proved to be short-sighted and counterproductive with the two clusters developing their Electronic Medical Records (EMR) systems independently. This prohibited a seamless exchange of health data when patients moved over from one cluster to another (Sinha et al., 2013). To resolve this problem, an interim solution known as the Electronic Medical Record Exchange (EMRX) was implemented in 2004 (Muttitt, 2011). Although this facilitated an exchange of health data between the two clusters, such exchanges were limited to a document-level exchange due to the lack of structured data (Sinha et al., 2013). This meant that the 'smartness' to facilitate clinical decisions, research and surveillance was technically infeasible. Another issue with the EMRX was that the data exchange was not governed by any regulation to protect the privacy of patients. These drawbacks posed new challenges which had not been anticipated earlier.

To compound the challenges mentioned above, yet another threat that made the healthcare transformation complicated stemmed from the realization that the movement of patients, especially of the chronically ill and the elderly, was not just going to be restricted to the two public healthcare clusters, but might actually span across the entire spectrum of primary, secondary, tertiary, intermediate and long-term healthcare, in the public, private as well as voluntary welfare healthcare sectors.

The technical and structural challenges apart, another critical issue said to confront Singapore is what is known as the 'silver tsunami'. The expression 'silver tsunami' refers to a rapidly ageing population, one of the key indicators of which is a population's median age. By 2030, 1 in 5 Singaporeans is estimated to be over the age of 65, and by 2050, Singapore is predicted to be one of the demographically oldest countries with a median age of 55 years (Institute of Policy Studies, 2012). About 85% of those aged 65 and above are forecasted to develop one or more chronic illness that would necessitate life-long treatment (Singapore Department of Statistics, 2010 as cited in Ng, Sy & Li, 2011). These projections, in addition to the already existing challenges in Singapore's healthcare system, led to the stark realization that a transformation of the existing system is what is needed. The rapid growth of the number of the elderly would also mean that the time that Singapore has at its disposal to transform its healthcare system is rather limited as the silver tsunami is likely to impact Singapore's healthcare landscape sooner rather than later. The situation would result in severe bed/resource crunches in hospitals, stretching the waiting time for a hospital bed to about 12 hours, thereby affecting productivity (Pang, R., personal interview, April 20, 2015; Chowdary, K.P.H., personal interview, May 15, 2015). The demands of such a rapidly ageing population could possibly render the currently hospital-centric healthcare model unsustainable (Gunapal et al., 2016).

The turn of events discussed above eventually led to an integration of the existing healthcare systems not only to better meet the demands of an ageing population, but also to acquire the capability of a seamless exchange of health information that would bring to fruition Singapore's strategic vision of '*One Patient One Record*'. Such a capability is considered critical to ensure that coordinated care is delivered in the most appropriate settings as well as to prevent any undue strain on the country's healthcare infrastructure (Ng et al., 2011).

Thus, over time, through steps and missteps, insights were gleaned and lessons were learnt that helped modify the Singapore healthcare system. The need for integrating healthcare provision beyond the public sector to include the private and voluntary welfare sectors was strongly felt at the political level. This led to a reorganization of Singapore's healthcare system. As a consequence, the healthcare system was split into six major clusters or RHS (Regional Health Systems) in 2010 (Gunapal et al., 2016) for better care coordination and integration to offer holistic, patient-centric healthcare. An independent agency called AIC (Agency for Integrated Care) which was formed in 2009 to develop the Long Term Care sector, was tasked with integrating these six clusters at the national level and help patients navigate the RHSs (Koh & Cheah, 2015). Figure 5.2 depicts Singapore's six health clusters.

Figure 5.2. Singapore's Regional Health Systems network

This reorganization is believed to facilitate a shift from an episodic level care to a patient level or patient-centric care (Shum & Lee, 2014). Each RHS is made responsible for the health of the population in its region, and this entails collaborations and partnerships beyond the public healthcare sector. RHSs have to collaborate with general practitioners (private, primary care sector) as well as voluntary welfare organizations to offer holistic and integrated care. Each RHS is anchored by an acute care hospital that partners with other healthcare providers in the region such as community hospitals, nursing homes and general practitioners[1].

The aforementioned challenges and changes culminated in the Ministry of Health (MOH) initiating in 2010 the nation-wide electronic health record systems referred to as the National Electronic Health Records (NEHR) (Chowdary, K.P.H, personal interview, May 15, 2015). To realize its vision of *'One Patient, One Record'*, the Singapore government made the NEHR system a key component of the Intelligent Nation 2015 (iN2015), a 10-year plan (Sinha et al., 2013). The NEHR program is estimated to break even within 7 to 8 years of its implementation, and generate a net present value of approximately S$581 million (Muttitt et al., 2012). MOH Holdings Pte Ltd, the holding company of Singapore's six RHSs, is tasked with the implementation of the NEHR. The NEHR is being developed based on international standards and its deployment is supported by an enterprise architecture (Accenture, 2012).

Considering that 80% of healthcare costs are generally incurred in the last few years of an individual's life time (Pang, R., personal interview, April 20, 2015),

---

[1] http://www.aic.sg/page.aspx?id=137

the focus of Singapore's healthcare industry is also geared towards providing better intermediate and long-term care for the elderly. It is also considered equally important to step up 'preventive healthcare' initiatives in preparation for an ageing population. This might mean ushering in 'Smart Homes' equipped with monitoring systems to enable management of chronic health conditions. Such 'Smart Homes' may be supported by Cloud, SaaS, and mobile technologies. Initiatives such as Smart Home and Smart Mat that are already under trial have the potential to ease this resource crunch faced by hospitals (Pang, R., personal interview, April 20, 2015; Chowdary, K.P.H., personal interview, May 15, 2015). The Smart Mat, which is being experimented with in the Changi Prison Medical Center, is a non-intrusive device that can capture breathing and heart rates, which are important parameters in clinical monitoring. In addition, it could also determine the quality of sleep, and send timely alerts e.g. fall alerts. Its flexible design allows it to be mounted on mattresses or cushions in home or clinical settings. All these 'Smart' initiatives would be linked to the NEHR to capture the data flows for monitoring purposes and for determining the appropriate level of care for a patient or consumer (Chowdary, K.P.H., personal interview, May 15, 2015).

5.3 NEHR Implementation

The NEHR was planned to be implemented in two phases (Muttitt et al., 2012). The targets set for phase 1 to be accomplished by the second quarter of 2012 were:

- one-way sharing of health data with limited integration partners,
- viewing of health information through the NEHR portal

And for phase 2, the targets to be accomplished by 2015 were:

- increased integration, bi-directional health data flows
- more information and data sources
- reconciliation services
- increased portal access

Singapore is regarded by informants and scholars alike to be making good progress on e-health with the NEHR currently in its second phase of implementation. The bigger plan is for the NEHR to be eventually migrated to H-Cloud, the consolidated healthcare cloud that would host mission critical systems for all public hospitals, specialty centers and polyclinics. The H-Cloud is facilitated by ST Electronics (Info Software) Systems and owned by Integrated Health Information Systems (IHiS), which is a 100% subsidiary of MOH Holdings Pte. Ltd. IHiS serves as MOH Holdings' project management wing for healthcare projects. The target for ST Electronics (Info Software) Systems is to make the mission critical systems in the various healthcare institutions interoperable by year 2017, and have these migrated to H-Cloud subsequently. As of May 2015, over 500 servers and over 200 applications of various healthcare institutions had already been migrated to H-Cloud. Although these initiatives were primarily focused on the public healthcare sector, it is hoped that the private and voluntary welfare organizations would also be consolidated under H-Cloud in due course (Chowdary, K.P.H., personal interview, May 15, 2015).

It is widely believed that embracing cloud technology will bode well for Singapore's healthcare transformation although it is an arduous path ahead. The various healthcare institutions in Singapore had been functioning like silos, having their own infrastructures, and running their own applications. With the advent of the NEHR, these silos needed to be made interoperable, consolidated, and migrated to H-Cloud (Chowdary, K.P.H., personal interview, May 15, 2015), which is a private cloud (Ng, 2013). Such a step understandably involved heavy initial investments. Although the Government subsidized these costs to some extent, the healthcare institutions also had to deploy their own funds for the purpose. There was some initial resistance to this initiative which was however overcome gradually when these healthcare institutions started seeing the bigger picture and were convinced of the long-term benefits of such a direction (Chowdary, K.P.H., personal interview, May 15, 2015). The present research speculated whether this initial resistance may have been due to the fact that interoperability increases transparency and reduces the information asymmetry that healthcare providers have been profiting from. Significantly, it has been acknowledged by no less a person than Mr. Khaw Boon Wan, the former Minister for Health (August 2004 to May 2011), that information asymmetry could be a valid reason for healthcare market failure not just in Singapore but around the world in general. Equally important, the disclosure by Mr. Khaw that competition in healthcare occurs more at the micro level between doctors or departments, than at the higher level of the clusters was revealing (Chang, 2010). A somewhat similar view was encountered during an expert interview when it was shared that the potential loss of business to other institutions due to reduced information asymmetry resulting from interoperability was not a major concern among the public sector healthcare institutions which were anyway facing a resource crunch and struggling to meet the healthcare needs of the public (Chowdary, K.P.H., personal interview, May 15, 2015). This raises the question of whether the reorganization of Singapore's public healthcare system into the six RHS clusters would suffice to infuse fairness and efficiency into the healthcare market leading to 'better service and lower cost for patients' as was hoped for when the two giant clusters of SingHealth and National Healthcare Group were first formed in 2000 (Chang, 2010), but proved to be unsuccessful later.

Despite the various steps taken to improve Singapore's healthcare system, there is still scope for further improvement. The insights extracted from the multiple slices of data used in this case study in general, and the expert interviews in particular, bring into focus some of these improvement possibilities. These are possibilities that emerge from the immense potential values that can be created and captured in a well-connected e-health ecosystem - knowledge that has come to light over the course of this study. What this would mean in the Singapore context is that there is further scope for more values to be created and captured if

110   *Capturing Value in Digital Health Eco-Systems*

only the health data is exploited so as to make the healthcare system in Singapore patient-centric or even population-centric, as advocated by Gunapal et al. (2016) and Nurjono, Valentijn, Bautista, Lim & Vrijhoef (2016). The remainder of this section is a discussion of these improvement possibilities:

- E-health in Singapore is thought to be primarily aimed at improving patient experience, increasing productivity in healthcare institutions, and promoting interoperability among disparate health information systems. It is however still lagging behind countries like the US and Japan in terms of providing Continuum of Care (Pang, R., personal interview, April 20, 2015). Healthcare Information Management Systems & Society (HIMSS), a global thought leader of healthcare transformation through technology, defines 'continuum of care' as a system's capability to track patients over time across an array of healthcare services at all levels and intensity of care (Young, Clark, Kansky & Pupo, 2014). One of the mechanisms that is believed to enable continuum of care is an integrated health system. The current model of healthcare in Singapore does not support care transition and the resulting continuum of care as there is no system in place for a transparent exchange of information that will enable tracking of GP clinics' patient referrals to hospitals. In a survey conducted by Accenture on healthcare IT adoption and health information exchange in the primary care sector, Singapore was found to be lagging behind several other countries (Accenture, 2012) in this respect. This shortcoming is also acknowledged by Nurjono et al., (2016) who claim that where the Singapore healthcare system is concerned there is some 'weakness' at the primary care level, the level at which the integration of the healthcare system begins. This shortcoming may be attributed to the fact that the NEHR is currently more focused on the public sector healthcare ecosystem financed and regulated by the Singapore government. As for the private sector healthcare institutions, the only group that is impacted, in the sense that they are encouraged to acquire access to the NEHR for a monthly subscription fee, is the General Practitioner (or GP or 'family doctor') clinics. As part of the GP IT enablement program launched in 2009, Project Clinic Electronic Medical Record and Operation System (CLEO) was created to facilitate access to the NEHR for the GP clinics as well as to enable an exchange of health information among them (Accenture, 2012). These clinics are generally said to be disgruntled with the monthly subscription model that they are currently subject to, for gaining access to the NEHR (Pang, R., personal interview, April 20, 2015). It is to be noted that as of early 2015, only 40% of the GP clinics have had access to the NEHR and that the government has been soliciting support from the remaining GP clinics (Singapore Parliamentary Report, February 12, 2015, Question No. 494). Apparently there still exists a perceived lack of correspondence between values created and values captured among the GP clinics. It may be pointed out that this is an instance of the participation dilemmas discussed in Chapter 3. It is not clear how these dilemmas will be resolved to the satisfaction of the GP clinics so as to make Singapore's dream of '*One Patient One Record*' come true. As pointed out by Chin (2000), the dream may not come true until the doctors perceive the need to embrace technology for clinical or financial benefits.

Further, if the other groups in the private sector such as the secondary, tertiary, intermediate and long term care providers are also linked to the NEHR, it may augment the efficiency and competitiveness of the healthcare system by enabling a comparison of performance between the two sectors. This is known as 'network effects' (Benkler, 2006) where nonproprietary production in the networked information economy leads to the sharing of efficiency and group sustainability. A case in point is the evidence cited from Denmark and Australia where private sector hospitals function alongside public sector hospitals, creating thereby a competitive environment that pressurizes both the sectors to improve their performance (Dash & Meredith, 2010). Of late, Singapore's public sector hospitals have come under scrutiny for serious lapses in their quality of care. A most recent example is the widely reported slips in Singapore's public hospitals – a wave of hepatitis C infections leading to deaths in Singapore General Hospital (YahooNews, 2015), and the threat of tuberculosis outbreak among pediatric patients in the National University Hospital (Khalik, 2015). Numerous studies have shown that carefully designing and implementing competition among hospitals belonging to both public and private sectors can drive innovation, quality of care and efficiency in the healthcare system (Lim, 2005; Dash & Meredith, 2010). A well-connected e-health ecosystem which fosters an exchange of information and promotes transparency in the system may be a suitable vehicle to accomplish this.

- The healthcare model in Singapore is 'fee-for-service' unlike the model in the US and some other countries which is 'pay-for-performance'. In a fee-for-service model, care quality may suffer because the providers are focused on providing as many services as possible to maximize their 'economies of scope'. It is a well-known fact that healthcare providers tend to exhibit entrepreneurial behavior since "every dollar spent on healthcare is a dollar earned by them" (Lim, 2005, p.464). In a pay-for-performance model, on the other hand, payments to providers are linked to their provision of quality care to patients. This model offers incentives to providers to focus on the patient instead of just providing unnecessary services. One of the clinical quality indicators that support the latter model is the readmission rate for patients within 30 days after discharge (Lim, 2004). In the US, for instance, the federal government imposes penalties on hospitals with higher than the benchmark readmission rates (Rau, 2014). In Singapore, this rate seems to have been on the rise for the public hospitals, going from 11.7% in 2011 to 12.2% in 2012 and 2013 as per the statistics shared by Minister for Health, Mr. Gan Kim Yang, in a Parliamentary response (Singapore Parliamentary Report October 7, 2014, Question No. 114) - a rate comparable to that in the

US. In this context one cannot help raising the question if such a comparison is reasonable considering that Singapore ranks among the most efficient healthcare systems in the world, while the US is ranked at the bottom among advanced economies. In a recent study conducted by Bloomberg, Singapore was ranked the most efficient healthcare system among a total of 51 countries studied, while the US was relegated to the 44th rank (Chen & Wong, 2014). In this context, it is both pertinent and important to consider how e-health could play a role in improving the above-mentioned quality indicator. Gunapal et al. (2016) drew attention to the fact that big data, i.e., health data residing with the six RHS clusters aggregated at the national level, can help predict as well as mitigate the risks leading to hospital readmissions. Another move that may improve this quality metric is promoting transparency by publishing the readmission rates of the various hospitals so that healthcare consumers can exercise their choice of providers based on the quality of care. It is interesting to note that a suggestion to publish such consumer-friendly information was made in the Parliament in October 2014 (Singapore Parliamentary Report, October 7, 2014, Question No. 114), which clearly signals the rise of healthcare consumerism in Singapore. This is another instance of how a well-connected e-health network has the capacity to reduce information asymmetry, promote transparency and make the healthcare system consumer-centric.

- The current model of the public healthcare system in Singapore, despite its strategic reorganization into the six RHS clusters, is still considered hospital-centric and unsustainable in the face of the ever-rising demands from an aging population (Gunapal et al., 2016). It is suggested that there should be a shift towards population health management and that information technology capabilities such as 'big data analytics' could be harnessed to be better able to manage population health. However, such a capability can be reached only if sharing of health information across the independent health systems is enabled so that all the data can subsequently be standardized and aggregated. It stands to reason that such aggregated public health data presents unprecedented opportunities to manage health at the population rather than the individual level. There are however some challenges of creating big data such as standardization of patient data coming from disparate systems, and sufficient anonymization of such data to prevent unauthorized use to discriminate in employment or insurance coverage (DeNardis, 2014). Confidentiality of such patient data is ensured in two ways – through an audit system to identify unauthorized entry and through stiff penalties for any IT security violations (Singapore Parliamentary Report (2012, March 6), News Highlights). A recent, noteworthy development on the 'big data' front is the steps taken by IHIS in this direction such as the establishment of systems like Electronic Health Intelligence System (eHINTS) and Business Information (BI) System to support big data analytics (Kelleher, 2015).

- The health cloud (H-Cloud) may already be underway, but there seems to be some ambiguity on the ground with regard to patient data residency and sovereignty (simply put, 'who owns what?'). These were said to be outstanding issues yet to be addressed and resolved to the satisfaction of the stakeholders (Pang, R., personal interview, April 20, 2015). A careful search for information to clarify these issues (purposive sampling) revealed that the custodian of patient data across all the six clusters is HSOR (Health Services & Outcomes Research), a department established in 2005 within NHG, is one of the six RHS clusters. As for data ownership, the clusters themselves are the owners of their respective patients' data (Gunapal et al., 2016). In its capacity as the custodian of patient data, the HSOR was tasked in 2014 with developing what is known as the RHS database (big data) with support from IHIS for population management related research activities (Gunapal et al., 2016). In fulfilling this task, HSOR adheres to strict data linking and anonymization protocols as established through the Personal Data Protection Act (PDPA).

5.4 Singapore's E-Health Ecosystem

In this section, the e-health ecosystem in Singapore is compared with the conceptual model developed in phase 1. Such a comparison is made for the purpose of noting similarities and dissimilarities, a knowledge of which would be essential to derive answers to the RQs as well as evolve the critical success factors for a sustainable e-health ecosystem.

5.4.1 Key Players

From the discussions above, it may be evident that the key players in Singapore's e-health ecosystem at this point of time are the regulators, patients, a subset of healthcare providers who are linked to the NEHR and an emerging infomediary. A concept map of Singapore's e-health ecosystem is presented in Figure 5.3.

The case study has also revealed that the health cloud (H-Cloud) is a prerequisite for a sustainable e-health ecosystem as the technology can scale to accommodate exponentially growing health data besides offering other benefits such as reducing providers' IT investment burden, supporting interoperability and decreasing information asymmetry, to name a few.

5.4.2 Values Created and Captured

At present, the ecosystem appears to be more provider-centric than consumer-centric. This means, the ecosystem in its present state offers rather limited potential for values to be created and captured as compared to what was envisaged in the conceptual model (Table 4.4). This is due to the fact that the ecosystem is still unfolding and hence the significant data flows discussed in section 4.1.5.1, 'E-Health Digital Data Flows: A Simplified Model' have yet to materialize.

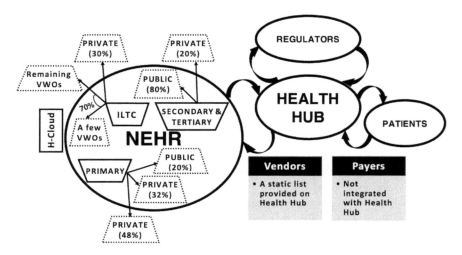

Figure 5.3. A concept map of Singapore's e-health ecosystem

However, considering the fact that initiatives such as www.healthhub.com are underway, it is hoped that, over a period of time, Singapore's ecosystem will evolve to be more inclusive of other key stakeholders and tilt the balance towards consumer-centrism. 'HealthHub' is intended to be a one-stop health information and services portal for citizens, offering them access to their medical records from the public healthcare institutions, as well as giving them access to health-related information and services (MOH Press Release, 2015). The ultimate goal of this venture is to promote health awareness among Singaporeans so that they are well-equipped to take responsibility for their health and wellness. This move is in alignment with Singapore's philosophy that healthcare is a shared responsibility (Stephanie, 2017) and that providing healthcare services upon demand is neither a viable nor a sustainable arrangement.

HealthHub has plans to extend the ecosystem through partnerships with organizations in the healthcare, wellness, recreation, fitness and food and beverage businesses, thus expanding the range of vendors in the ecosystem to promote consumer choice. Moreover, HealthHub offers consumers the opportunity to create values in the ecosystem through co-creation of content by sharing or contributing useful health and wellness information that may benefit other consumers in the ecosystem. Patient-entered health data for which the NEHR is mandated to make provision, is yet another opportunity for consumers to co-create data.

It appears that HealthHub has the requisite characteristics to play the role of the infomediary described in the conceptual e-health model, albeit with a narrower scope than envisaged in the conceptual model described in Chapter 4. The ecosystem seems more focused on connecting the public sector healthcare providers. Although HealthHub provides a static listing of the private sector healthcare providers, many of these providers are not linked to the NEHR. The exception to this is the primary care category in which about 40% of the GP clinics are currently linked to the NEHR. Similarly, payers are not well-integrated into the ecosystem, especially those such as private insurers and employers. As for the vendors, there is a static listing of various types of vendors such as pharmacies, laboratories and screening centres, but transactions with them are not yet enabled. The HealthHub is no doubt a positive step towards boosting consumer awareness of the choices available to them. It, however, limits their empowerment to make informed healthcare related decisions inasmuch as it provides information that is neither complete nor transparent.

5.4.3 The Ecosystem through a Game Theory Lens

It may be worthwhile to examine Singapore's e-health ecosystem using the game theory lens, which provides the conceptual framework for examining the ecosystem. The ecosystem presents instances of the participation as well as cooperation dilemmas discussed in Chapter 3.

5.4.3.1 Participation Dilemmas.

The GP clinics which are private sector healthcare providers evince symptoms of participation dilemmas. This group of stakeholders seems reluctant to participate in the ecosystem because they are required to pay a monthly subscription fee to access the NEHR. Obviously, this reluctance stems from the perception that the value they can capture from being part of the ecosystem, may not be worth the financial commitment they have to bear. This is clearly a manifestation of the 'productivity paradox', a participation dilemma. Yet another participation dilemma, namely the 'tragedy of the digital commons' is also observed to be at play in this context. Notwithstanding the fact that the GP clinics have to pay for access to the NEHR, they are also expected to contribute their patients' health data to the NEHR to support continuum of care. Such data as they contribute to the NEHR may be utilized by other providers in the network resulting in 'free riding'. It is the belief of some experts that an interoperable system based on the Continuum of Care Document Architecture (CCDA) devoid of any incentives to stakeholders will not work. The system should provide for some incentives to the key players to cooperate and share information, even though it may not necessarily be in economic terms. The participation dilemmas are thus the result of a perceived lack of fairness in the ecosystem.

5.4.3.2 Cooperation Dilemmas.

The ecosystem also exhibits cooperation dilemmas in the sense that it is not an inclusive ecosystem - it excludes the private sector healthcare providers from categories other than primary care. Yet, it is the private sector that provides numerous value-added services so much so that the Government has co-opted the

GP clinics as well as some private hospitals into the public system. For now, these excluded private healthcare providers do not have access to the NEHR. This is reminiscent of the 'narrow network', a cooperation dilemma. A narrow network limits information sharing and value-capture to a select group of stakeholders, which may threaten optimal utilization of health data in terms of its exchange and reuse beyond the network. The implication is that Singapore's vision of 'one patient one record' and 'continuum of care' may be endangered when patients move across the spectrum of public and private sector healthcare providers with a view to meeting their healthcare needs. Yet another implication is that a narrow network limits information available to the consumers which in turn does not help reduce 'information asymmetry', a barrier to consumer-centrism, and a threat to efficiency in the ecosystem. It is logical that the ecosystem should evolve to include other key players including private sector providers, payers and vendors. Without the participation of these players, sufficient values may not be generated in the ecosystem to keep the stakeholders engaged, and the benefits of e-health may not be harnessed in totality. The ecosystem should also evolve to add more capabilities aimed at reducing information asymmetry – such as enabling healthcare consumers to compare products and services on the basis of quality, price, and other relevant factors so as to be able to make informed decisions. HealthHub, the infomediary, may also benefit from the process since the value it captures will be proportionate to the values captured by the other players in the ecosystem. Correspondingly, the more an infomediary creates values that benefit the other players, the higher are the values it can capture from the ecosystem.

5.4.4   Critical Success Factors for a Sustainable Patient-Centric E-Health Ecosystem

The central posit in this monograph is that the values derived by the key players in the ecosystem should be sufficient to justify their participation in the ecosystem for its balance and sustainability. In other words, the key players should have the opportunity to capture values in proportion to the values they create, without which they will not be incented to participate fully.

Some propositions put forth by the experts for a sustainable ecosystem, were:

- To encourage participation from general practitioners (private sector), a more effective method to get them to leverage the NEHR would be to move from the monthly subscription model to a pay-per-use model. A two-way model may be more optimal where healthcare providers such as GP clinics who pay for access to health information in the NEHR also get incentivized in some manner for contributing information to the NEHR. There are many revenue sharing models for this ranging from a fixed, one-time offset of subscription fees for each unit of data, to a recurring royalty-credit whenever that unit of data is used by others.
- Healthcare institutions may also consider charging patients nominally (like $1 per healthcare episode) for creating/maintaining their records.

Within-Case Analysis of Singapore's NEHR   **117**

- To augment participation from the general public and patients in the process of co-creating health data, PHR systems that can link to the NEHR may be sponsored by insurers, for example. Insurers may go a step further and incentivize their policy-holders to use the PHR system which will, in turn, facilitate preventive healthcare versus the less-than-ideal diagnostic healthcare which increased costs for the insurers. It may be noted that an e-health ecosystem actually meets the criteria for what economists term a two-sided market - selling healthcare products and services to patients on one side and market data to other players on the other.
- It will benefit Singapore's citizens if the Government fully sponsors the Next Generation Nationwide Broadband Network (NGNBN) which facilitates Smart Homes and Smart Mat. As a payoff on a national scale, the regulators and policy makers can gain access to more health data captured by these monitoring devices and conveyed to the NEHR. It may be noted that about 80% of the NGNBN infrastructure is already in place, and the network is expected to be fully operational by 2025.
- There is potential for new business opportunities for healthcare providers, hospitals, in particular. They may set up call centers to monitor smart homes to be able to direct healthcare resources to where they are most needed.
- Homes may be made 'smart' through NGNBN, smart phones and a base kit (like the one for cable TV) which will enable the public to subscribe to healthcare services they require and pay for these based on their level of consumption ('pay as you use').

5.5 Lessons Learnt

- It was observed during the expert interviews that the informants took care to be politically correct. Due to such a tendency on the part of the informants, uncovering issues in Singapore's e-health ecosystem took considerable probing as well as time. To ease the informants' predicament, they were apprised of the opportunity to review the transcription of their interviews before giving their final approval for its inclusion in the data.
- Literature on Singapore's e-health initiative was found to be far limited compared to similar literature available for other advanced countries like the US. Therefore, a variety of other secondary data sources including news websites, industry journals, industry reports, blogs and parliamentary proceedings, had to be relied upon to piece together the case study.
- The Singapore case study also led to the refinement of the semi-structured interview for the US case study. For example, a question on cloud technology was added to ensure that related data is captured. 'Cloud technology' was a naturally emergent topic for the Singapore informants who seemed to be up to date with the H-Cloud initiative. Considering that cloud technology is not a priority in the US within the context of debates centered around

118   *Capturing Value in Digital Health Eco-Systems*

the Affordable Health-care Act (Obama Care), care was taken to explicitly incorporate a question to ensure data collection on the topic.

5.6 Conclusions

Based on the insights gathered from the case study, it is reasonable to conclude that Singapore's e-health initiative comprising the three key components of NEHR, H-Cloud and HealthHub is making progress towards sustainability. It is envisaged that the H-Cloud will be accomplished by year 2017, making the mission critical systems in the various public healthcare organizations interoperable. This would be a significant game-changer for Singapore's healthcare industry. For example, the GP and neighborhood clinics may no longer have the autonomy associated with possessing a patient's medical history. Alliances and chains may emerge to exploit the added values from such open digital flows.

However, it is hard to ignore that the current initiative is focused on the public healthcare sector which makes up 80% of secondary and tertiary care, and 20% of primary care. 80% of primary care is provisioned by the 2000-odd GP clinics, and as of early 2015, only about 40% of them could have access to the NEHR. Getting the remaining 60% on board the NEHR may be challenging owing to the participation and cooperation dilemmas discussed earlier. Some of these dilemmas were also corroborated by the experts interviewed. Whether the Government will follow the carrot (incentives) and stick (penalties) approach to get them on board is not yet clear. And whether other key players such as the private healthcare providers in sectors other than primary care and payers such as private insurers and employers will be eventually co-opted into the ecosystem is as yet an open question. It has long been suggested that the government should be consistent in its approach towards all the providers. This would mean, among other things, exposing the public sector healthcare providers to market forces and have them compete with the private sector providers for their fair share of patients (Lim, 2005). This goal is achievable through a well-connected ecosystem complete with the entire spectrum of public and private healthcare providers. It is to be noted that the manner in which the public sector hospitals mostly engage their private sector counterparts through the utilization of the latter's excess bed capacity by leasing their beds (Singapore Parliamentary Report, March 6, 2012, News Highlights). The patients using these leased beds are however only treated by their respective public hospital doctors. Thus, the two sectors hardly have opportunities to cooperate or compete on parameters that might matter to a healthcare consumer. If that happens, it could create a climate of healthy competition between the two sectors.

In conclusion, although the findings from the case study show that Singapore's e-health ecosystem currently falls short of the conceptual model in some respects, the country is still undeniably one of the very few countries that have made giant strides in e-health. This may be evident from the fact that it recently received the prestigious DataCloud Enterprise Cloud Award for H-Cloud squashing competition from hundreds of other global submissions (Lee, 2015). Given Singapore's quest for excellence in providing public services, the country's e-health ecosystem hold out a promise to evolve into a truly consumer-/citizen-centric system. To quote Porter (2010), "Value should always be defined around the customer, and in an efficient health care system, the creation of value for patients should determine the rewards for all other actors in the system" (p. 2477).

5.7 Chapter Summary and Recap

Singapore enjoys the rare distinction of being ranked one of the healthiest countries in the world, as well as, credited with having one of the most efficient healthcare systems in the world. The country's healthcare transformation was necessitated by the looming 'silver tsunami' (a rapidly ageing population), a phenomenon which was predicted to render the country's predominantly hospital-centric healthcare model unsustainable in the mid to long run. This was one of the major push factors that led to a reorganization of the country's healthcare system in 2010 into six RHSs which were each made responsible for the population's health in their respective regions. To complement this, a nation-wide electronic health record system (NEHR) was also initiated in the same year. The grand plan envisaged was to migrate the NEHR to a cloud platform referred to as the H-Cloud, by 2017. Also underway is HealthHub, a one-stop health information and services portal for citizens, which includes an interface for citizens to access their health records on NEHR. All these moves are believed to gradually facilitate a shift towards a population-centric healthcare. However, at this point in time, the ecosystem falls short of the conceptual model in terms of both the set of key players and the range of values that can be created and captured by them. This is due to the fact that the ecosystem is heavily focused on the public healthcare sector and only selectively inclusive of the private healthcare sector. The ecosystem is also found to manifest the participation and cooperation dilemmas discussed in sections 3.2.4 and 3.2.5, which pose a challenge in terms of getting the private sector primary healthcare providers on board. A truly patient-centric e-health ecosystem like the one conceptualized in phase 1 of this research can materialize only if the entire spectrum of public as well as private healthcare providers participate and cooperate in the e-health network. Without this happening, Singapore's dream of One Patient One Record will be hard to realize. Nevertheless, as of date, Singapore still remains one of those few countries that have made significant progress in e-health.

The next chapter presents the within-case analysis and report of the second case study, the US' HITECH program.

CHAPTER

6

Within-Case Analysis of the US' Hitech

This chapter presents the second of the case studies undertaken as part of phase 2 of this research, namely the US' HITECH program. As said earlier, phase 2 of this research is focused on validating the conceptual model of a patient-centric e-health ecosystem that was developed in phase 1. The chapter begins with an introduction to the US healthcare system, goes on to chart the country's healthcare transformation journey and proceeds to contrast the country's e-health ecosystem with the conceptual model developed in phase 1. This is done with a view to evolving a set of critical success factors for patient-centricity and sustainability of the country's e-health ecosystem. The chapter ends with a short account of lessons learnt in the process of conducting the case study.

6.1 Introduction

The US has always been at the forefront of research and innovation in medical technology, often ushering in successful, new-generation interventions (Shi & Singh, 2015). It makes substantial investments in medical research in partnership with the National Institutes of Health, possesses possibly the best medical workforce trained in best-of-breed medical schools and hospitals, and has one of the most advanced healthcare systems in the world (Vitalari, 2015). As is often the case, advances made in science and technology tend to create a demand for new products and services. This is especially so in respect of healthcare industry. As a matter of fact, competition among American hospitals is often a matter of how modern and sophisticated the equipment and gadgets in their possession are. Needless to say, such a trend triggers an overuse of technology primarily to ensure that the huge capital investments made on technological equipment are worthwhile and profitable. So, on the one hand, there are medical professionals who are eager to put their latest equipment and gadgets to use and, on the other, there are patients who are led to believe that these latest technologies provide better outcomes (Shi & Singh, 2015). Thus, other than being a world leader in medical technology, the US also happens to have the most expensive as well as the most sophisticated healthcare system in the world.

The US tops the list of industrialized nations in terms of its annual healthcare expenditure with 17.1% of its GDP spent in 2014 (Global Health Observatory, 2016). This roughly translates to $2.87 trillion (Squires & Anderson, 2015). In 2000, the Institute of Medicine issued an impactful report titled 'To Err is Human', which pointed out that between 44,000 and 98,000 deaths occurred in US hospitals as a result of medical errors and that half of these deaths were actually preventable. Prevention would not only have saved precious lives, but money too to the tune of $17-$29 billion per year. More recently, it has been estimated that the US would have an additional trillion dollars at its disposal, if only it maintained its spending on healthcare at the same percentage of GDP as the next highest spending country in the world (Fuchs, 2014). Furthermore, it is projected that if the current state of affairs continued, the US spending on healthcare will rise to $5.4 trillion by 2024, which will amount to 20% of the country's GDP (McCarthy, 2015). This is cause for concern as such a rate of government spending on healthcare alone, might compromise spending on other areas including security and well-being, thereby putting at stake the overall welfare of the country (Britnell, 2015).

The US healthcare system also happens to be one of the most complex systems in the world involving an extensive array of interrelationships among care providers and payers (Moses III, Matheson, Dorsey, George, Sadoff & Yoshimura, 2013). Although healthcare facilities such as hospitals, clinics, doctors' offices and other facilities are owned by both private and public entities (Cummings, 2015), in terms of both healthcare provision and financing, the US is more at the private end of the public-private sector mix. A breakdown of the US hospitals by ownership types is shown in Figure 6.1.

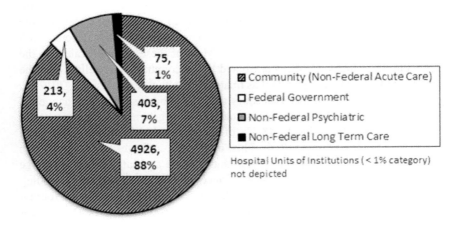

Figure 6.1. Number of US hospitals by ownership type (Total: 5627)
Source: AHA Hospital Statistics, 2016

A further breakdown of the community hospitals by ownership type is shown in Figure 6.2.

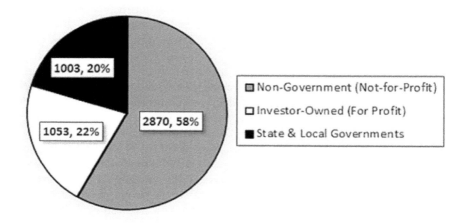

Figure 6.2. Community hospitals by ownership type (Total: 4926)
Source: AHA Hospital Statistics, 2016

Paradoxically, the US also happens to be the only industrialized nation that does not guarantee universal health coverage for its people (Davis, Stremikis, Schoen & Squires, 2014), despite the heavy spending on healthcare that almost cripples its economy. As many as 33 million people (10.4%) were uninsured in 2014 (Smith & Medalia, 2015), the reasons being not wanting insurance due to its growing unaffordability to its rejection by private insurers on account of pre-existing conditions (Cummings, 2015). The burden of healthcare costs on the population has been so heavy that it was estimated to contribute to about 3 in 5 bankruptcies in 2013 (Lamontagne, 2013). Most of these woes prevalent in the US healthcare industry are often attributed to the country's over-reliance on the private sector for delivering as well as financing healthcare (Goldsmith, 2012).

However, over the last few decades there have been significant changes in the sources of healthcare spending in the US, with general government health expenditure gradually rising to 48.3% of the total health expenditure in 2014 (Global Health Observatory, 2016). This, in a sense, indicates the government's declining reliance on the private sector, and may be attributed to a wide range of government insurance programs made accessible to the Americans including federal programs like Medicare, Medicaid, Children's Health Insurance Program (CHIP), and health plans offered by individual states as well as by the Department of Veteran Affairs. Nevertheless, the 48.3% government spending on healthcare is still considered far lower than the average for most other developed countries which approximately stands at 72% (Britnell, 2015). More than half (55.4%) of the private healthcare spending was contributed by employer-based insurance systems in 2014 (Smith & Medalia, 2015). Such a trend is not only a huge financial liability for businesses, but also results in a misallocation of time for corporate leaders, which may lead to decreased productivity and weakened competitiveness. A case in point is the bankruptcy of General Motors in 2009 due in part to its health insurance liabilities which it had to pack into the price of the car, and, in turn, ended up losing its competitiveness in the automobile market (Herzlinger, 2010).

A key barrier to universal health coverage in the US is the fragmented state of its healthcare system (Vitalari, 2015) which is also the prime reason for the country's astronomical healthcare spending. The system comprises a wide range of healthcare delivery, insurance and payment mechanisms that may be financed publicly as well as privately, and lacks the presence of a central governing agency to coordinate and integrate its services (Shi & Singh, 2015). According to the 2011 estimates of the Institute of Medicine, roughly one third of the total healthcare expenditure was wasteful spending on account of various factors such as inefficient and redundant services, fraud, abuse, steep prices and administration costs, and, last but not least, missed opportunities for prevention (Britnell, 2015). It is thus apparent that the US healthcare system is a poor value proposition relative to its cost (Herzlinger, 2010). The system has reached a tipping point when, undoubtedly, a change is much needed, although how the change can be accomplished is not clear (Sharfstein, Fontanarosa & Bauchner, 2013).

6.2 The US's Healthcare Transformation Journey

A significant step in the US' long pursuit to transform the healthcare industry was the Health Information Technology for Economic and Clinical Health Act (HITECH Act) which was signed into law by President Obama in early 2009 (Rouse, 2014). The HITECH Act was a component of the American Recovery and Reinvestment Act of 2009 (ARRA) economic stimulus bill created to stimulate the adoption of EHRs and supporting technologies. The Office of the National Coordinator for Health Information Technology (ONC or ONCHIT), a division within the US Department of Health and Human Services, was assigned the responsibility of creating a strategic plan for a country-wide interoperable health information system. The HITECH Act made provision for incentives worth $27 billion to spur adoption of EHRs among physicians and hospitals (Blumenthal & Tavenner, 2010). The legislation also compelled hospital administrators to enforce policies to protect the confidentiality of patient data in compliance with the Health Insurance Portability and Accountability Act (HIPAA) Privacy Rule of 2003 and Security Rule of 2007 (Ilie, Van Slyke, Parikh & Courtney, 2009; Clarke III, Flaherty, Hollis & Tomallo, 2009). A further legislation called the HIPAA Omnibus Rule was enacted in 2013 to direct extensive changes to the HIPAA Privacy and Security Rules and strengthen them so as to conform to the guidelines of HITECH (Rouse, 2014). The HIPAA rules also applied to the IT vendors collaborating with healthcare providers in order to prevent them from commercially exploiting the patient data in their possession (Blumenthal, 2009). According to Accenture (2012), the HITECH Act was a strategic shift in the stance of the government which now wanted to take charge of the out-of-control healthcare *industry* by driving IT adoption and facilitating greater connectivity across the fragmented system in the process.

In early 2010, the ONC awarded $548 million worth of grants to the 50 US states, the District of Columbia, and the five federal territories, through the State Health Information Exchange (HIE) Cooperative Agreement Program. The grants were funded through the HITECH Act for the purpose of developing infrastructure and augmenting capability to exchange patient data among the various healthcare organizations both within and across the states (The Office of the National Coordinator, 2010). A commendable feature of the Act was that the state or the state-designated agency which received the award was to be held responsible for developing the plan and tracking performance for reporting as per the guidelines in the award notice (SearchHealthIT, 2010). Every healthcare provider eligible for the EHR Incentive Program was required to be part of at least one such HIE.

Another major development following the HITECH Act of 2009 was the Patient Protection and Affordable Care Act (also known as ObamaCare or simply ACA) which was voted into law in March 2010 after a long, contentious political battle (Blank, 2012). The ACA was aimed at increasing health insurance coverage for Americans and also at controlling rising healthcare costs (Mangan, 2015). The Act specifically accentuated the significant role that health IT could play in meeting the healthcare related goals of quality and efficiency (Buntin, Burke, Hoaglin & Blumenthal, 2011). A key component of the ACA was the Health Insurance Exchange Marketplace which opened in October 2013. It is an online marketplace for health insurance that enables shoppers to select the best plan in accordance with their needs by making use of the platform's capability to display a side-by-side comparison of different health plans. Estimates suggest that this marketplace, which operates without a broker, will be able to provide affordable insurance to up to 29 million people by 2019 (ObamaCare Facts, 2016). All these recent legislations and their provisions are believed to have the ability to improve the performance of the US healthcare system over time on dimensions of access, efficiency and equity.

6.3 HITECH Act Implementation

In recognition of the fact that the widespread adoption of EHRs is inevitable to integrate the fragmented healthcare industry in the US, the federal government made an unprecedented commitment of $27 billion dollars towards the cause through the 2009 HITECH Act (Blumenthal & Travenner, 2010). This amount was to be spent over a span of 10 years on incentives to physicians and hospitals to encourage their adoption and meaningful use of EHRs. The incentive scheme was structured as a carrot and stick approach – carrot by way of additional payments if

Within-Case Analysis of the US' Hitech **125**

providers demonstrated meaningful use of EHRs, and stick in terms of cut-backs - if it was not demonstrated (Adler-Milstein & Bates, 2010). The 'meaningful use criteria' evolved over three stages from 2011 to 2016. Stage I focused on data capture and sharing, stage II on using advanced clinical processes, and stage 3 on improved outcomes particularly, better population health (HealthIT.gov, 2013). Key features of meaningful use include electronic prescribing, quality measurement reporting and information exchange (Torda, Han & Scholle, 2010). While an EHR may have several functionalities, three of these in particular are considered critical to promoting meaningful use, because of their potential to improve quality of care and reduce costs (Menachemi & Collum, 2011). These core functionalities are:

- Clinical Decision Support (CDS) tools that support a physician in making evidence-backed decisions regarding patient care. Widespread use of this system is believed to help reduce medical errors and improve quality of care.
- Computerized Physician Order Entry (CPOE) systems that facilitate computerized orders for medical products and services such as drugs, laboratory tests, and radiology. This system is intended to prevent medical errors that are made in the process of making sense of a physician's handwriting.
- Health Information Exchange (HIE) that enables secure and real time sharing of patient data with other healthcare providers, pharmacies, emergency departments and so on. This has the potential to reduce costs by eliminating redundant tests and improves efficiencies.

Thus, the focal point of this reform was a patient-centred EHR. A 2010 tracking of national trends in technology used by Pew Internet & American Life revealed that as many as 3 in 4 adults used the Internet, the top reason being, to search for health-related information (Ahern, Woods, Lightowler, Finley & Houston, 2011). The rise of consumerism abetted by technology was becoming evident, and this was what was addressed by the reform. The message for the healthcare providers through this reform was that they had to adapt to a climate of doing more with less (Orlikoff & Totten, 2010).

In spite of the fact that more than 90% of the US hospitals used computers for some purpose or the other as of 2008, only about 9% of these hospitals had comprehensive health IT systems with a basic EHR. Around the same period, 13% of the primary care offices had fully functional EHR systems (Hammond, Bailey, Boucher, Spohr & Whitaker, 2010). However, within a span of about eight years, by 2014, about 75% of the non-federal acute care hospitals had adopted at least a basic EHR system with clinician notes, and one-third had adopted a comprehensive EHR (Charles, Gabriel & Searcy, 2015). As for the physicians, more than 80% had adopted basic EHRs and more than a third, fully functional ones by 2014 (Shay, 2016). The latest update on the EHR incentive program is that nearly $32 billion worth of incentives have been paid out to more than 484,000 healthcare providers as of the end of January 2016 (Roberta, 2016). This goes to show that the HITECH incentives have played a crucial role in catalyzing EHR adoption and meaningful use among physicians and hospitals alike.

6.3.1 Challenges in Implementing E-Health

Adler-Milstein & Bates (2010) highlight some key challenges encountered by hospitals in their EHR implementations. One of these was the heavy upfront expenses which in the case of some hospitals were higher than their single largest capital expenditure over a five-year period. Other challenges included uncertain ROI and the transient negative impact of implementing the technology, such as loss of productivity, staff downtime and possibly loss of revenues too in the process of adapting to it before it became a way of life. Yet another significant barrier was of course resistance from the physicians hired by these hospitals. The challenges faced by primary care practitioners in implementing EHR systems were nearly the same (Fleming, Culler, McCorkle, Becker & Ballard, 2011), albeit on a smaller scale. The experts interviewed attested to the above-mentioned challenges and also brought to attention the fact that resistance to use EHRs was particularly high among the older generation of physicians. In fact the physicians' reluctance to use the EHR platform has been turned into a new business model by some vendors who started offering scribe services to capture physician notes in the EHR and generate reports for physicians to sign off, all for the purpose of satisfying compliance (Raja, N., personal communication, August 23, 2015). It appears that sometimes the facts on the ground may be different from official rhetoric. Even to date there are some medical practices that are so resistant to technology intrusion into their treatment rooms that they do not mind being left behind in terms of access to tools for effective care-giving (Raja, N., personal communication, August 23, 2015). Moreover, these physicians also do not mind paying penalties for non-adoption as they consider this a cheaper alternative as compared to the long term costs associated with maintaining and upgrading the EHR (Katragadda, M., personal communication, September 2, 2015). Some small, rural groups also refrain from investing in EHR fearing the investment and maintenance costs in spite of the federal incentives (Raja, N., personal communication, August 23, 2015). After all, the government incentives go only as far as to encourage adoption and not beyond. Some physicians and hospitals do not want to invest in the EHR as yet, as the technology itself is still evolving. Stories abound of failed EHR implementations which have made them wary, have resulted in their decision to wait to catch the maturity curve much later (Raja, N., personal communication, August 23, 2015).

The patient-centric stance of the reform was instrumental in bringing about a gradual but steady paradigm shift in healthcare which required that healthcare systems move from focusing on episodic care to focusing on integrated care. This meant placing the patients at the center of care and focusing not only on their illness, but also on their wellness (Serbanati et al., 2011). A healthcare system that only addressed episodic care was considered as falling short of this key requirement. This paradigm shift set the direction for the healthcare systems of the future - it became imperative for them to be interoperable so as to facilitate a sharing of patients' data both within and outside the provider setting, thus paving the way for all health data pertaining to them to be captured in a longitudinal record (Paun et al., 2011). This was what was considered to make for a truly patient-centric experience.

The EHR incentives do not simply hinge on its adoption. Rather, for meaningful use, the healthcare providers are required to collaborate with other stakeholders in the region or community to participate in Health Information Exchanges (HIE) which facilitate a sharing of health information according to nationally recognized standards for better care coordination and quality of care (Lassetter, 2010). The grand plan is for these regional HIEs to be connected to their state-level HIEs which, in turn, will be tied in to form a nation-wide health information exchange known as eHealth Exchange (McCann, 2014). The eHealth Exchange actually started off as an ONC federal initiative in 2006 and was transitioned in 2012 to The Sequoia Project, a private sector initiative, for support. The eHealth Exchange currently claims to be the largest HIE network in the US, supporting more than 100 million patients, 4 federal agencies, nearly 50% of US hospitals, 26,000 medical groups, 3400+ dialysis centers and 8300 pharmacies across all the 50 states (The Sequoia Project, 2017). The allure for the participating organizations is the opportunity to do away with expensive customizations to interface with trading partners and to reduce their legal fees by leveraging common standards, legal agreements and governance accessible through the eHealth Exchange. In return, the participating organizations pay two types of fees to the eHealth Exchange, namely the annual eHealth Exchange Network Participation Fees which is a sliding scale based on annual revenues, and Testing Fees, when they first come onboard and/or seek to validate their product. This is how eHealth Exchange, the player whose role resembles that of the infomediary in the conceptual model, has structured the network as a mutually beneficial arrangement. While eHealth Exchange is a national level infomediary, the regional HIEs that it connects through the network are smaller infomediaries or sub-networks which in turn connect a fewer number of healthcare organizations at the level of a region or community.

The federal support for HIEs was for a short term and limited to providing these entities with start-up funding. It was however left to the states to discover a sustainable business model for the long haul either on their own or in partnership with the private sector organizations (Adler-Milstein, Bates & Jha, 2011). The end objective for the federal government was to consolidate the state level HIEs to a national level network by developing appropriate technical standards. This will facilitate aggregation of patient data leading to creation of 'big data' for population health management activities which can possibly improve healthcare outcomes both clinically as well as fiscally (Kayyali et al., 2013). At the time of the reform, the states already had entities known as Regional Health Information Organizations (RHIO) which possessed the capability to facilitate clinical data exchange at the local level within defined geographical areas. The RHIOs typically connected healthcare providers, payers, laboratories and public health departments. The financial incentives tied to healthcare providers demonstrating meaningful use through health information exchanges bolstered interest in the RHIOs which already had the structure to achieve HIE. Thus, the erstwhile RHIOs became HIEs after the reform. It is to be noted that HIE may have two meanings depending on the context. The term may refer to the process of exchanging health information or to the entity facilitating such an exchange.

Setting up the HIEs and getting the stakeholders to meaningfully exchange data in ways that would result in improved efficiency and quality outcomes was a challenge. For one thing, the manner in which the stakeholders chose to exchange data was determined based on self-interest rather than on the larger public interest of achieving efficiency and quality in healthcare. This resulted in a narrow set of transactions taking place through the exchange in contrast to the vast potential of the HIE for broader based and meaningful transactions (Adler-Milstein et al., 2011). Some HIE business models were not sustainable because the issue of misaligned incentives among its stakeholders could not be resolved satisfactorily. For instance, providers and patients were unwilling to pay for transactions from which the payers benefited the most financially (Adler-Milstein et al., 2011). Another challenge was that some of the HIEs within a geographical area might be directly competing with each other and therefore least inclined to share their patients' data with the other HIEs (Romeo, 2013). Apart from the challenges discussed above, there were some more in the form of disparate standards being adopted by the HIEs that deterred interoperability and consequently exchange of data, among them. Thus the very building blocks of the national network, the HIEs, remain siloed and disconnected.

The mixed results yielded by e-health implementations such as EHR and HIE across the US led to the realization that e-health involved more than just technical design (DesRoches et al., 2010). It was observed that in many instances the technology was implemented just for its own sake without any foresight of the goals to be accomplished through the technology (Mettler & Eurich, 2012). Needless to say, such implementations turned out to be failures. Thus, the need for a carefully planned implementation strategy involving the key stakeholders was recognized. For the success and sustainability of e-health implementations, a business model to collaborate and create value with concerned stakeholders was deemed a necessary component of the strategy (Van Limburg, Gemert-Pijnen, Nijland, Ossebaard, Hendrix & Sevdel, 2011). Mettler and Eurich (2012) suggest aligning comprehensive knowledge about technology's potential with business acumen and sensitivity to customer needs to create a business model that would be both economically and socially sustainable.

6.3.2 Existing Business Models

Thus, seven years into the HITECH reform, the quest for viable business models that will help to achieve a fully integrated, sustainable national health information

Within-Case Analysis of the US' Hitech **129**

network continues. In undertaking this search for a sustainable business model, it may be pertinent to examine some business models that are already in use in the much fragmented healthcare industry of the US.

6.3.2.1 Managed Care Organizations (MCO)

Various business models to deliver healthcare have been in use in the US for a few decades now. One such model that started gaining popularity in the 1970s particularly among employers weighed down by the cost of insurance for their employees was the Managed Care Organizations (MCO). Typically financed by the employer or government, the MCO still remains the most dominant healthcare delivery system in the US (Shi & Singh, 2015). It functions in this way: an employer or government negotiates a contract with an MCO to offer a selected health plan for their employees as an alternative to purchasing expensive insurance plans. Three types of managed plans are offered by an MCO, namely (i) Health Maintenance Organizations (HMO), (ii) Preferred Provider Organizations (PPO) and (iii) Point of Service (POS) (Katragadda, M., personal communication, September 2, 2015). In an HMO, the primary care provider acts as a gatekeeper to coordinate care, and the plan only pays for care within the network. A PPO on the other hand offers a wider choice of healthcare providers and does not have the primary care provider acting as a gatekeeper for enabling access to specialist care. Moreover, it is more flexible when compared to an HMO in that it even allows members to seek care outside of the network. Understandably, a PPO is more expensive and it pays less for care received outside the network as different from care received within the network – a strategy intended to capture value. A POS lets the member choose between an HMO and a PPO at the point of care. An MCO is a capitation contract which functions somewhat like a fixed price contract, where the financial risk is transferred to the providers. The providers receive a fixed annual or monthly payment per member regardless of how much care is provided to the member during the period. If less than expected care is provided, they profit, and, conversely, if more than expected care is given, the difference will be absorbed by them. Such an arrangement benefits the payers because of the certainty it offers in terms of budget. However, it may be cause for concern if providers compromised on the quality of care to economize (Frakt & Mayes, 2012). The major shortcoming of this model was its flawed design of rewarding physicians for not giving adequate care to their patients (McLean, 2007). In other words, the physicians had no accountability for the health outcomes of their patients, which is a disadvantage.

6.3.2.2 Accountable Care Organizations (ACO)

Around October 2011, a new value-based model of healthcare known as Accountable Care Organizations (ACO) was introduced by the government through the Medicare Shared Savings Program (Fisher, McClellan & Safran, 2011). An ACO comprises doctors, hospitals and other healthcare providers who collaborate to ensure that the chronically ill Medicare patients have timely access to quality care provided in an efficient manner. It is to be noted that Medicare is a national social insurance program that covers Americans aged 65 years and older who have contributed to the program during their employment. In addition, it also covers younger people with certain disabilities and conditions. It is one of the biggest sources of insurance in the US and covered more than 50 million Americans as of 2014 (Smith & Medalia, 2015). As of February 2016, there are a total of 448 ACOs (Centers for Medicare and Medicaid Services, 2016). An ACO differs from an MCO in the sense that it is designed to address the shortcomings of the latter. In an ACO arrangement the financial risk is split between the providers and payers, and, additionally, bonus payments are tied to providers meeting quality parameters. Moreover, the ACO does not confine members to its network for all the care they require. Such a model is believed to prove efficient as well as sustainable especially with the advent of the EHR whose capabilities can support this model in terms of keeping track of physician's interventions and associated outcomes (Frakt, 2015). The ACO model has caught on even outside of the Medicare program with the private sector insurers and providers entering into similar contracts (Fisher et al., 2011; Frakt, 2015).

6.3.2.3 Pay for Performance (P4P) Systems

An increasingly acclaimed payment model currently implemented by the federal government in its Medicare program is the Pay for Performance (P4P) system to control healthcare costs without any compromises on the quality of care. P4P systems are devised to measure the performance of the providers along selected dimensions using pre-defined indicators. These measurements are used to assess them on the basis of their efficiency so as to compensate them accordingly (Cromwell, Trisolini, Pope, Mitchell & Greenwald, 2011). P4P is a move away from the traditional fee-for-service model which incentivizes providers for providing as many services as possible without any regard for cost or efficiency. Such a system not only contributes to an increase in healthcare costs but also results in inefficiencies as a result of overutilization of scarce resources. In line with the patient-centric standpoint of the reform, the P4P system also offers scope to embed preventive care within the healthcare system by rewarding providers on the basis of how well they maximize preventive care to improve health outcomes. The fee-for-service system, on the contrary, rewards providers for neglecting preventive care which is cheaper though more efficient than diagnostic care. An anomaly within the system is that providers stand to profit from their patients' adversity - the worse off a patient's condition, the more the provider can profit. For example, a provider will earn more if a diabetic patient suffers kidney failure and less if the patient's condition is kept under control through preventive care such as routine checks (Montgomery, 2016). Although the P4P system shows promise to better align the interests of the providers and the patients, the question remains whether it will be impactful and to what extent it can improve health outcomes.

6.3.2.4 Provider-Led Healthcare Networks

A rising trend that has been apparent in the US healthcare industry since the enactment of the ACA is provider-led healthcare networks. Big hospitals vertically integrate by merging with medical practices and solo physicians who cannot afford to invest in EHR and comply with regulatory reporting requirements so as to form large healthcare networks (Zinberg, 2016). Moreover, they also offer health plans in just the way as insurance companies do, to give their patients the added benefits of convenience and care (Accenture Healthcare Consumer Survey, 2012). These provider systems may build their own plan, or acquire an existing plan, or partner with a health plan (Eggbeer, 2015).

One such network is Sutter Health, a not-for-profit healthcare provider with presence in Northern California, Oregon and Hawaii. One of the biggest healthcare systems in the US, it includes clinics and community healthcare providers, and generates $9 billion in revenues. The clinics and community providers in its network are required to pay a subscription fee to have access to Sutter Health's EHR platform. The federal incentives currently cover the subscription fees. However, when the incentives are phased out, the healthcare providers in Sutter Health's network will have to bear the subscription and support costs on their own. Some of these providers attempt to recover their EHR-related costs by charging their patients for access to the system at their end, and some others just absorb the costs. Sutter Health also has an insurance arm that offers health plans akin to those offered by managed care organizations (MCO) – HMO, PPO and POS. While Sutter Health is a not-for-profit organization, its insurance arm is for-profit (Katragadda, M., personal communication, September 2, 2015).

6.3.2.5 Personal Health Records (PHR)

Another business model that garnered attention around the time of the reform was the Patient Health Records (PHR). The concept of PHR managed to arouse interest by positioning the technology as being patient-oriented as different from the provider-oriented nature of the EHRs (Sunyaev, Chornyi, Mauro & Krcmar, 2010). Several PHR models emerged in the market, the most notable ones being GoogleHealth and Microsoft HealthVault which to date have only had limited success in spite of riding on big and established names. Inexplicably, GoogleHealth exited from the service in early 2012.

6.3.2.6 Integrated Delivery Systems (IDS)

There are also other healthcare delivery models simply referred to as integrated delivery systems (IDS) which are essentially various forms of strategic partnerships among hospitals, physicians and insurers. An IDS can be described as a network of organizations that either directly provides or arranges to provide a well-coordinated continuum of care to its members, a defined population, and assumes accountability for the health outcomes of the population both in clinical and fiscal terms (Shi & Singh, 2015).

6.4 The US' E-Health Ecosystem

This section discusses the US' e-health ecosystem and how it compares with the conceptual model developed in phase 1. The RQs articulated in Chapter 1 provide the framework for this discussion. Similarities and dissimilarities are highlighted, based on an analysis of which, a set of critical success factors for the ecosystem to be patient-centric and sustainable is identified.

6.4.1 Key Players

A review of the characteristics of the US healthcare industry, the recent (disruptive) reforms, their implications, and the myriad business models that exist, all confirm the highly complex, pluralistic nature of the industry. A concept map of the US e-health ecosystem is shown in Figure 6.3.

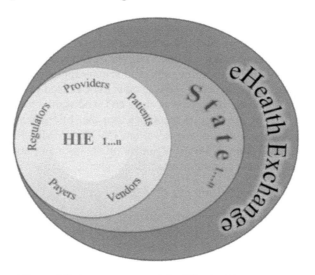

Figure 6.3. A concept map of the US e-health ecosystem

The ecosystem to be established through eHealth Exchange at the national level is still in a state of flux and is expected to keep evolving for quite some time to come. The US e-health ecosystem has the same players as those identified by the conceptual model, the only difference being the players' complex inter-relationships. The players in the US e-health ecosystem are in a highly entangled and layered web of various types of business model arrangements at various levels. It will be apt to state that the national level health information network will be an ecosystem of ecosystems – a high level one that will connect the smaller ecosystems of HIEs. The HIEs, in turn, connect several entities that include patients, providers (hospitals, medical practices as well as independent physicians), payers, provider-payer arrangements such as MCO, ACO, IDS,

*Within-Case Analysis of the US' Hitech*133

provider-led networks, laboratories, vendors and public health departments. The HIE is in itself a complex web, and it will eventually have to be connected to the other HIEs to form the national network. The ecosystem will also include regulators at multiple levels – the federal government, the state government and the state-designated organizations. Needless to say, bringing such a large-scale e-health ecosystem to fruition from its current fragmented state is a mammoth task, but it may be a necessary step to transform the industry.

6.4.2 Values Created and Captured

The ecosystem, as it is, allows limited scope for the range of values in the conceptual model to be created and captured in entirety. However, subsets of these values are observed in pockets, at various levels in the ecosystem. For all these values to be available in the aggregate on a single platform, eHealth Exchange should evolve to unify the smaller units of HIEs.

6.4.3 The Ecosystem through a Game Theory Lens

Viewed through the lens of game theory, it is apparent that e-health in the US is a non-cooperative game. One of the conditions that characterizes a non-cooperative game is that the benefits derived from cooperation are not returned to the members in a manner they consider equitable (Ford, Wells & Bailey, 2004). The stability and sustainability of the ecosystem are threatened by the two types of dilemmas discussed in Chapter 3, sections 3.2.4 and 3.2.5, namely participation and cooperation dilemmas.

6.4.3.1 Participation Dilemmas

The US e-health ecosystem does not offer much scope for participation dilemmas that arise on account of a 'productivity paradox' largely because of the federal mandate in the form of the HITECH Act of 2009. The Act made provision for the EHR Incentive Program which encouraged EHR adoption and meaningful use among healthcare providers through incentives. Currently 75% of the US hospitals and more than 80% of the medical practices have basic EHR systems, and more than one-third in each group have advanced EHR systems. However, there are pockets of healthcare providers who have not invested in an EHR despite the federal incentives because they find technology intrusive and disruptive, and because they believe that paying penalties for non-adoption will be less costly than investing in EHR and incurring long term costs on its maintenance.

The other participation dilemma, namely the 'tragedy of the digital commons' is even more prevalent than the previously mentioned dilemma. One of the criteria for a meaningful use of the EHR is to share patient data with other healthcare providers and entities such as laboratories, and, public health departments, to ensure coordinated care for patients. Many healthcare providers are however unwilling to share their patients' data with other stakeholders. The medico-legal considerations aside, this could be because of their perception that the other stakeholders are free-riders on the assets they painstakingly create. A considerable effort goes into converting a patient's paper records into electronic health records – manual data entry, scanning of documents, and verification of data and documents before filing (Kumar, N., personal communication, August 23, 2015; Katragadda, M., personal communication, September 2, 2015). Understandably, the healthcare providers are disinclined to allow others to benefit from their effort for free.

6.4.3.2 Cooperation Dilemmas

HIEs at the regional, state and federal levels continue to be developed in a siloed manner owing to two key reasons: lack of interoperability resulting from incompatible technical standards, and competition with the other HIEs in the geographical region (Romeo, 2013). HIEs may not be open to sharing health information with other HIEs that are perceived to be their direct competitors. If only HIEs were willing to exchange data with others (HIEs), the patients in their networks would greatly benefit, as this would inevitably lead to the creation of longitudinal health records for patients, regardless of where they receive care. However, from an HIE's perspective, this may adversely cut into its profitability, as patients now have an option to endlessly switch from HIE to HIE without incurring any additional costs. Not surprisingly, an HIE is reluctant to share patient data with other competitors, which in effect makes it a 'silo' or a 'narrow network'. It follows then that healthcare systems prefer to penalize patients switching between networks by imposing on them an additional cost in some form. For instance, a healthcare system may not prefer to share their patients' data with another system, but refusing to provide patient data when needed is not an option they enjoy. Therefore, the healthcare system may follow a policy to release patient data in a format of their choice (paper or electronic form) as a result of which patients are denied the advantages of a longitudinal health record. Much worse, they may even charge a fee from the patients for that service (Katragadda, M., personal communication, September 2, 2015). Recently, EPIC, one of the leading EHR vendors in the US was accused of designing its systems to be 'closed records' which are not interoperable with other systems (Tahir, 2014). Incidentally, healthcare providers may also be willing to cut down on their service fee in exchange for a health plan which is designed to discourage patients from seeking care outside the network with the result 'narrow networks' (Landman, 2015) begin to emerge.

A 'narrow' network results in 'information asymmetry' which renders patients immobile and incapable of accessing as much information as they may wish to in order to exercise informed options and be free to switch between networks, if needed. Consequently, such networks are deemed to be less efficient and less consumer-centric, as they deliberately deny or limit consumer mobility or freedom which is crucial in making informed choices.

As discussed above, the dilemmas, when applied to and interpreted in the context of the US healthcare industry, provide a deep insight into the key issues afflicting the industry, which quite obviously stand in the way of accomplishing a nationwide health information network.

6.4.4 Critical Success Factors for a Sustainable Patient-Centric E-Health Ecosystem

What follows is a brief discussion of some recommendations made by the experts for stabilizing and sustaining the US e-health ecosystem:

- The benefits accruing from optimizing the use of the EHR will eventually outweigh the burden of investing in it, and the subsequent maintenance costs it entails. The EHR empowers the hospitals and physicians by providing access to a connected healthcare ecosystem through which they can gain further insights into diseases and cures, leading to a more effective and better-coordinated care. By the time meaningful use stage 3 criteria are met, it is hoped, the healthcare providers will have achieved efficiencies that translate to increased earnings which, in turn, would help cover their EHR maintenance costs in total. Among other things, using the EHR would help reduce the length of an appointment possibly from 45 minutes to 15 minutes, thus saving a physician's valuable time. The time saved could be used for more patient consultations, which in turn helps generate more revenues. Moreover, being part of an ecosystem also expands the network and the market for the healthcare providers and opens doors for new business opportunities.
- Smaller medical practices often find EHR investments forbidding, and therefore join larger healthcare networks which have already made the investment. For a subscription fee, they can access the EHR and benefit from its use, and they may recover the amount by collecting a small fee from their patients. What the patients get in return is an empowerment to create, own and access their own EHR accounts as well as the ability to decide who they share the information with.
- Private insurers have the power to ignite e-health adoption by making it mandatory and paying for it through compensatory mechanisms similar to what the federal government does through the EHR Incentive Program. In return, they gain efficiencies that will help them detect and prevent fraud, waste and abuse by healthcare providers, which in turn would result in substantial savings for them over the long run. Recent developments like EHR-facilitated e-visits are also of benefit to insurers who pay much less for their health plan holders' e-visits than what they may have to pay for their face-to-face consultations with their physicians.
- Patients can also play a role in driving e-health by insisting that their providers have EHRs; this would inevitably lead to the creation of a longitudinal health record which will be ubiquitous and portable. After all, patients have the right to demand access to information which is their own and for which they pay by way of consultation fees, charges for lab tests, and all costs including medications.

- The government (federal or state) can mandate a filing back of patient data into a government owned system, and act as the custodian of their citizens' health information. The 'big data' thus accumulated may be used not only for formulating effective health policies, but also for developing early warning and advisory mechanisms, leading to an overall improvement in population health. Availability of big data can also foster analytics and informatics opportunities which the pharmaceutical companies can use to find cures and arrest outbreaks.

  In view of the massive stores of data such a system may have to accommodate, cloud-based solutions which could be an excellent option both technically and economically feasible. Although there are no major players who are currently using the cloud technology, given the exponential rate at which health information is growing, it is highly likely that in about ten years, almost all health information may be on cloud. In the near future cloud technology will be an integral component of e-health programs, as several facets of e-health depend on the evolution of the technology for their materialization. However, there are also concerns about how the technology can ensure security and confidentiality of highly sensitive health information. Although encryption technologies could help mitigate such concerns, nothing may possibly work better than the government assuming responsibility for the security and confidentiality of the e-health ecosystem in order to maintain public confidence – similar to what the government has been doing for the national financial system by provisioning the FDIC (Federal Deposit Insurance Corporation), an independent agency which insures deposits and monitors financial institutions to ensure consumer protection.

6.5 Lessons Learnt

- There was an abundance of literature available for the US case study and this presented a challenge in terms of obtaining a balanced view and staying on course.
- The US healthcare industry is excessively fragmented and populated with countless healthcare delivery models. The case study could only touch upon some of the dominant models. Needless to say, any situation can be turned into an opportunity and a business model by resourceful vendors. For instance, HIT vendors commonly offer scribe services to physicians who are not inclined to use the EHR but nevertheless have to comply with the meaningful use criteria in order to be eligible for the EHR incentives. These HIT vendors require the physicians to register an account with them, download a software from their website and install it in a computer in their consultation room. The physicians also have to share their appointment calendar with the HIT vendor so the vendor can have a scribe waiting virtually to take notes real time while the physician attends to the patient. What comes as a surprise is that these scribes themselves may be qualified physicians based in other

Within-Case Analysis of the US' Hitech 137

countries like India. It seems like the EHR Incentive Program has provided a second career to physicians in developing countries, whether or not it has resulted in meaningful use of the EHR.

6.6 Synthesis of Case Findings

A meaningful transformation has long eluded the US healthcare industry possibly on account of the continued dissent among the political leadership and the other key stakeholders about 'how' such a transformation could be brought about. Several reforms have been proposed in the past, though nothing substantial seems to have been accomplished - with the possible exception of two landmark reforms - the HITECH Act of 2009 and the Patient Protection and Affordable Care Act (ACA) of 2010. Although it is too early to conclude if these reforms have succeeded and, in what measure, they have nevertheless resulted in some sweeping changes to the US healthcare industry's landscape.

It is said that, compared to other developed countries, the US healthcare system has been consistently underperforming on measures of health outcomes (Davis et al., 2014). Many experts expressed the view that a digitization of patient health records combined with an effective use of information technology would go a long way in controlling the spiraling healthcare costs and the deteriorating quality of care that plague the US healthcare industry (Bandyopadhyay et al., 2012). The government therefore decided that any reform of the healthcare industry should actually begin with establishing a national-level health information technology infrastructure. This resulted in the creation of the EHR Incentive Program which is intended to support healthcare providers in their transition to IT and to promote a meaningful use of the technology to improve healthcare outcomes. The program was successful to the point that it could spur a rapid EHR adoption among hospitals and physicians. The government also provided grants to the states and territories to develop Health Information Exchanges (HIE) that would gradually pave the way for a seamless exchange of patient data within as well as across their jurisdictions. One of the criteria for a meaningful use of the EHR stipulates that the healthcare providers share patient data with other entities including other healthcare providers and laboratories when required, as such a step would improve patient care in terms of coordination and quality. This criterion encouraged healthcare providers to be part of an HIE so they can have access to a platform through which they can exchange patient data with other entities and demonstrate meaningful use. However, even within an HIE, it was ensured that data exchanges were limited to transactions that benefitted the transacting members. Thus the proposition of HIE only met with mixed results owing to two key reasons. First, there is a lack of interoperability because of the absence of a will to impose common technical standards to exchange data. Second, there is a misalignment of incentives between the investors in the technology and the actual beneficiaries from the technology. These factors, if not resolved, will result in gross underutilization of the vast potential of the HIEs in respect of improving health outcomes. It is time that appropriate business models were evolved that could create sufficient values for all stakeholders of HIEs, failing which their economic sustainability would be at stake. It stands to reason that the lofty ideal of a national health information network (eHealth Exchange) can be achieved only when the HIEs, the building blocks, are in a position to sustain themselves.

6.7 Chapter Summary and Recap

The US, a world leader in medical technology, ironically also happens to be home to one of the least efficient healthcare systems in the world. The country's mammoth, unsustainable spending on healthcare has triggered several key healthcare reforms to put the country's healthcare system on the path for a major overhaul. One such reform was the HITECH Act (2009) which made provision for incentives for adoption of EHRs by physicians and hospitals. The incentive scheme adopted a carrot and stick approach to motivate EHR adoption – additional incentives if meaningful use of EHRs was demonstrated and cutbacks if the criterion was not met. The ultimate goal of this reform was to forge connectivity across the country's fragmented healthcare system in the hope that it would lead to efficiency and consequently, a drop in the country's healthcare expenditure. Although the reform no doubt spurred EHR adoption, its intended goal of connectivity has not been fully realized owing to the participation and cooperation dilemmas discussed earlier, which only confirms the major barriers to e-health identified early on in this research. A set of critical success factors has been proposed to overcome these dilemmas and make the ecosystem both patient-centric and sustainable.

The next chapter (Chapter 7) synthesizes the results of the Singapore and the US case studies through a cross-case analysis with a view to mobilizing new knowledge from these studies. Such new knowledge as elicited by comparing and contrasting these case studies would be utilized to modify the conceptual framework which would then be compared with literature to pursue a theoretical integration. This would further lead on to the task of addressing RQ2, namely educing the critical success factors that contribute to a patient-centric, sustainable e-health ecosystem.

CHAPTER

7

Cross-case Analysis and Findings

This chapter addresses the third and final phase of this research, which is aimed at extracting insights that will help address RQ2. It may be recollected that RQ2 was aimed at deriving the critical success factors for a sustainable e-health ecosystem. As part of this chapter, a cross-case analysis of the Singapore and the US case studies is conducted and reported. The knowledge acquired through the individual within-case analyses is used to compare and contrast the cases, and produce accumulative knowledge. The new knowledge acquired in the process is used to modify the conceptual framework, the implications of which are discussed, and the modified theory compared with the literature for theoretical integration.

7.1 E-Health in Singapore and the United States

Each ecosystem's journey towards e-health is unique. Several factors such as the size of the country, structure of its healthcare system, political climate, socioeconomic factors and even culture may influence the journey. The key imperatives leading to the initiation of e-health to transform the healthcare ecosystem may also vary from country to country. For instance, one of the significant factors that started Singapore on its healthcare transformation journey was the 'silver tsunami'. In the US, on the other hand, the key factor driving the country's healthcare transformation journey was the skyrocketing healthcare expenditure. Regardless of the unique circumstances that lead a nation to pursue e-health in order to positively transform its healthcare industry, the underlying concerns with respect to healthcare cannot be dissimilar across healthcare systems, countries or even continents. When it comes to healthcare, each ecosystem gravitates towards the common goals of improving quality, access and equity for its citizens, and e-health is adopted for its promise to meet these goals.

Scholarly literature has recognized e-health as a promising strategy that can transform the healthcare ecosystem by integrating it and facilitating the flow of patient data in ways that can enhance the effectiveness and efficiency of healthcare (cf. Hill & Powell, 2009). It in fact presents an opportunity for global standards and best practices in order to create a common framework for sharing and comparing health information so as to have the ability to "innovate on the edges instead of having to reinvent wheels" (Stansfield, 2008). However, facilitating such an exchange of health information has always been a struggle for want of an effective model to do so (Vest, Campion Jr. & Kaushal, 2013).

The challenges involved in implementing e-health on a national scale are often underestimated. These challenges are not just limited to technical issues, but in fact have more to do with political and economic issues (Blank, 2012; Yaraghi, 2015). It is not easy to predict or resolve all of these issues without any reference to lessons learnt from similar initiatives. It is therefore worthwhile to learn from others' successes and failures in this context. In view of this objective, it will be beneficial to develop a theoretical framework that would help conceptualize the issues with regard to the adoption and purposeful use of e-health as well as identify the critical success factors for a patient-centric, sustainable e-health ecosystem. Such a framework will particularly benefit e-health implementations in developing countries which cannot afford the luxury of failed experiments. It is hoped that in developing such a framework, RQ2 will be addressed. A cross-case analysis of the Singapore and the US case studies, it is hoped, will shed light on how e-health may be successfully implemented and sustained.

7.2 Health Statistics: Singapore vs the United States

It may be recalled that the basis of selection of the two cases, namely Singapore's NEHR and the US' HITECH program were the purposeful sampling strategies of 'maximum variation' and 'information richness'. While Singapore is considered a high-performing healthcare system, the US, on the contrary, ranks as a low-performing system, making these cases diverse. However, one common trait these countries share is their propensity to quickly embrace innovation and technology in areas they perceive would benefit their nations. Thus, both Singapore and the US are among the pioneers to experiment with e-health technologies, which offers scope for this research to learn from these two countries' experiences.

It may be helpful for the cross-case analysis to begin with some most recent health statistics available for the two countries from the World Health Organization (WHO) as shown in Table 7.1.

Based on 2014 statistics, Singapore is classified as a healthy country with a life expectancy of 83 years (Global Health Observatory, 2016). In 2014, the life expectancy for the US was significantly lower at 79 years despite the fact that the country's per capita expenditure more than doubled Singapore's (after purchasing power parity adjustment) in the same year. In a sense, the proportion of the US expenditure on healthcare relative to GDP is nearly four times as much as Singapore's.

Singapore's healthcare system is "universal and unique" in the sense that it effectively combines free market principles with government control (Meng-Kin, 1998). Singapore adopts a mixed public-private healthcare delivery and financing

Cross-case Analysis and Findings **141**

Table 7.1. WHO statistics on healthcare indicators for 2014

Indicators	Singapore	The US
Life expectancy at birth (years)	83	79
Total health expenditure as % of gross domestic product	4.9	17.1
General government health expenditure as % of general government expenditure	14.2	21.3
General government health expenditure as % of total health expenditure	41.7	48.3
Private health expenditure as a % of total health expenditure	58.3	51.7
Out of pocket expenditure as a % of total health expenditure	54.8	11.1
Out of pocket expenditure as a % of private health expenditure	94.1	21.4
Total health expenditure per capita in US$	2752	9403
Out of pocket expenditure per capita in US$	1508	1044
Total health expenditure per capita in Int$ (PPP*)	4047	9403
Out of pocket expenditure per capita in Int$ (PPP*)	2218	1044

*PPP – Purchasing Power Parity – An economic theory that estimates the amount of adjustment needed on the exchange rate between countries in order for the exchange to be equivalent to each currency's purchasing power.

system that is based on personal responsibility as well as social solidarity (Britnell, 2015). This, however, results in patients having to pay heavy out-of-pocket costs which make up nearly 55% of the country's total healthcare expenditure. In contrast, the US does not offer a healthcare system that is universal and well-regulated (Moses III et al., 2013). Nevertheless, the out-of-pocket expenditure incurred by its people is much less i.e. 11% of the total healthcare expenditure. If the countries' out-of-pocket expenditure on healthcare per capita is compared after adjusting for purchasing power parity, it emerges that such spending in Singapore is nearly twice as much as in the US. The US government's expenditure on healthcare as a percentage of total healthcare expenditure is 48% which, though lower than the average for advanced economies, is higher than that of Singapore at about 42%. This discussion on the healthcare expenditure patterns in the two countries is pertinent inasmuch as it puts into perspective the key challenges confronting these healthcare systems, and how e-health can play a role in addressing these issues.

7.3 Similarities and Differences across Cases

Figure 7.1 shows the unique as well as common themes that surfaced from the two case studies, while bringing into focus both corroborating and contradicting evidence in the field. In the Venn diagram, the overlap lists themes or features that are apparent in both the ecosystems. Given the disparities in size and socio-

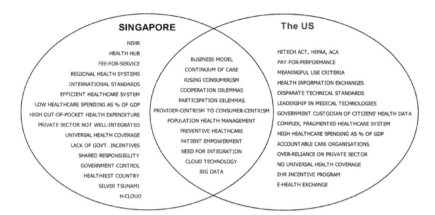

Figure 7.1. Unique and common themes from the Singapore and US case studies

political structure, such similarities across the two cases suggest a universal set of characteristics. The non-overlapping areas list the unique themes or features. Both the similarities and differences between the two ecosystems are discussed in this section.

As said earlier, Singapore and the US are both technologically advanced, and were among the earliest countries to initiate reforms in their healthcare systems through e-health technologies. While Singapore's major concern was about meeting the rising healthcare demands of its ageing population, the US was primarily concerned about bringing under control the country's astronomical healthcare spending, which, left unchecked, would be unsustainable (Herzlinger, 2010). Regardless of their differing motivations for their e-health journeys, their intended destinations are common – an integrated and well-connected patient-centric e-health network that will facilitate an exchange and reuse of health data (Neupert, 2009). It may be pointed out that the benefits that accrue from a capability to exchange and reuse health data have already been discussed in earlier chapters. A well-connected e-health network should, most importantly, be able to support a continuum of care and enable the creation of a longitudinal health record for patients. Having ready and ubiquitous access to their lifelong story of treatments, tests and prescriptions not only empowers healthcare consumers, but also supports their healthcare providers with regard to making the best possible decisions relating to their healthcare. However, as noted in the literature and field study, accomplishing such a patient-centric network is not a trivial undertaking. It requires, among other things, the cooperation of several stakeholders, especially the healthcare providers. It may be reiterated that EHRs are the building blocks of e-health, and healthcare providers are the stakeholders who are well-poised to contribute them. However, securing the cooperation of healthcare providers in adopting EHR technologies still remains a challenge. This challenge was observed to be at play in the case of Singapore as well as the US. The problem inherent in

Cross-case Analysis and Findings **143**

the challenge is that, as a rule, healthcare providers tend to perceive the new technology as a threat to their autonomy, and are skeptical whether the projected returns would be in keeping with their investments. Their question is whether it is worthwhile investing in a technology that may put them at a disadvantage; the value they capture may not be in proportion to the value they create in terms of their investments. They may even find it hard to accept the fact that the benefits of their investments may also be shared by stakeholders whose contribution to the common cause is minimal, and who are privileged to access the technology for free. Needless to say, such a perception of misaligned incentives stems from the reduced information asymmetry and the tragedy of the digital commons resulting from EHR investments. But do healthcare providers have a choice? The next section may help answer this question.

7.4 Optimal Ecosystem: A Game Theoretic View

A simple game theoretic analysis may help put the healthcare providers' predicament in perspective. It needs to be mentioned in this context that Sharma & Bhattacharya (2013) have developed some of the classic dilemmas faced in knowledge-sharing environments using game theory principles. This monograph adapts two of these dilemmas for the purpose of illustration.

For ease and convenience of analysis, the players in the e-health ecosystem may be classified into two groups namely, producers and consumers of health data as discussed in section 4.1.5.1, 'E-Health Digital Data Flows: A Simplified Model'. Health data is primarily produced by healthcare providers and consumed in various ways by the other players in the ecosystem such as healthcare consumers, payers, vendors, regulators and infomediaries. In this analysis, X denotes the producers of health data namely, the healthcare providers, and Y denotes the consumers of health data namely, the other players in the ecosystem. Using mathematical notations the game can be described in the following manner:

$$s = \text{number of strategies available to X}$$

Three major strategies namely

S_1 = non-adoption of e-health (may mean no HIT investment or HIT investment for intra-enterprise benefits)
S_2 = adoption of e-health in a narrow network
S_3 = adoption of e-health in an open network,

can be considered to be available for the healthcare providers; therefore $s = 3$.

S_x = strategy profile for X, expressed as $\{S_1, S_2, S_3\}$
P = payoff derived for each strategy of X
P_x = payoff profile for X, expressed as $\{X_1, X_2, X_3\}$
P_Y = payoff profile for Y, expressed as $\{Y_1, Y_2, Y_3\}$

Figure 7.2 shows the payoff graph for X and Y.

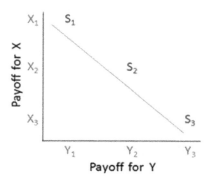

Figure 7.2. Payoff graph for healthcare provider strategies

The misalignment of incentives between the healthcare providers and the other players in the e-health ecosystem, a consequence of non-cooperative games, is evident from the payoff graph. The graph shows that the more restrictive the provider's strategy is, the less are the benefits for the other players.

However, the vast benefits of e-health cannot be forgone simply because e-health threatens the autonomy that healthcare providers have traditionally enjoyed. Moreover, it cannot be overlooked that healthcare consumers who are the ultimate beneficiaries of e-health are also the very source of the rich data in the possession of the healthcare providers. Hafen, Kossmann & Brand (2014) deplore the fact that healthcare consumers who are the source of health and health-related data have little or no control over such data, let alone their benefitting much from it. So, it is only fair that any value in e-health be defined around healthcare consumers. How much value the other players create for the healthcare consumers should be the basis of the decision on the reward payable to them (Porter, 2010).

Returning to the question whether healthcare providers have the choice to shirk EHR investments, the answer is they do not, if they are mandated by the regulator to invest. But even without such a mandate from the government, they may have no choice in this matter in view of the rising healthcare consumerism, especially in the advanced markets. As a matter of fact, consumers are now said to be gaining ground in healthcare, thanks to Internet trends and pro-consumer health policies (Guest & Quincy, 2013). However, this trend is corroborated by evidence from the Singapore and the US case studies. Healthcare consumers have come to expect the qualities they value in non-healthcare settings to be present in healthcare settings as well (Cordina, Kumar & Moss, 2015). Cordina et al. (2015) argue on the strength of the findings from their study that healthcare is no different from other industries from a consumer's perspective - 'customer satisfaction' is a common expectation that needs to be met by healthcare and non-healthcare organizations alike. It is imperative, therefore, that healthcare providers rise to their consumers' demand for ubiquitous access to their health information maintained in a longitudinal record. This rising trend of healthcare consumerism

Cross-case Analysis and Findings **145**

has a lesson for medical professionals as well; "the message to doctors is clear: Get Online – Do Not Be Left Behind" (Chin, 2000, p. 426).

A game theoretic analysis may also be helpful to evaluate how healthcare providers might choose to respond to rising healthcare consumerism. Consider a healthcare market served by two healthcare providers - A and B. Let the benchmark case be a situation where neither of the healthcare providers has adopted e-health. Adopting e-health entails costs in the form of EHR investments for the healthcare providers. Provider A serves a population size of "x" and has a profit that may be quantified as P. Provider B serves a population size of " y" and has a profit that may be quantified as Q.

$$A = \text{Provider 1}$$
$$B = \text{Provider 2}$$
$$x = \text{population served by A}$$
$$y = \text{population served by B}$$
$$P = \text{A's profit in the benchmark case}$$
$$Q = \text{B's profit in the benchmark case}$$

If A adopts e-health and B shirks e-health, or if A adopts e-health ahead of B, then x and P increase while y and Q decrease. This is because of the first mover advantage that A gains to create value for its patients in the form of electronic health records. Likewise, if B adopts e-health and A shirks e-health, or if B adopts e-health ahead of A, y and Q increase while x and P decrease.

If both providers adopt e-health around the same time, they may be able to maintain status quo in terms of the size of the population they serve. However, their profitability will now be lower as compared to the benchmark case due to their e-health-related investment and maintenance costs.

It may, therefore, be argued that the best strategy for both the providers would be to shirk investing in e-health to preserve their original profit as in the benchmark case. However, there is more to the analysis. First, A may not be aware of B's strategy and vice versa, as is the case in non-cooperative games. If A moves first, then B loses its patients to A, and if B moves first, A loses its patients to B. Of course, loss of patients comes with loss of profits (values) as well. Since there is an uncertainty regarding the other player's strategy, the best strategy for both players may be to play it safe by investing in e-health at least to maintain the status quo in terms of the size of the population they serve. Figure 7.3 shows the payoff matrix for the e-health adoption game.

Even if both the providers were to jointly decide not to adopt e-health in order to preserve their current profit ('P' amount of profit for A and 'Q' amount of profit for B as in the benchmark case), there is no guarantee that their patients will remain with them. The present-day healthcare consumers have better awareness of their options and may, therefore, tend to gravitate towards options that offer them more value. For instance, when healthcare consumers venture to go on medical tourism to countries where the healthcare system offers better value for their money (McLean, 2007; Herzlinger, 2010), they may not hesitate to switch their

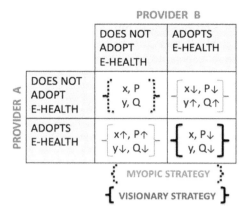

Figure 7.3. Payoff matrix for e-health adoption game

loyalty to a provider in an alternative healthcare market within their community or region, who offers more value (in this context value refers to e-health). This will endanger the status quo for A and B both in terms of their profits, and the population sizes they serve.

Thus, in any case, the best strategy for A and B would be to adopt e-health so as to be able to offer the best value possible to their customers. Moreover, embracing e-health may place them on the path to discovering new business opportunities and new markets. For both A and B, shirking e-health may seem a good strategy initially, but it may not be very long before they realize that it is not. The strategy is 'myopic' in the sense it is short-sighted and may prove detrimental in the long run. As against this, both A and B investing in e-health so as not to be left behind in terms of the opportunities brought forth by e-health may be said to adopt the best long-term strategy, referred to by this research as a 'visionary' strategy.

The foregoing discussion makes it clear, therefore, that under the present circumstances healthcare providers may not have any other sustainable option except to invest in EHRs lest they should otherwise be left far behind in the e-health game. It is likely that the e-health ecosystem may not be Pareto-efficient from the providers' perspective as it appears to make them worse off than status quo. However, with the prospect or possibility of innovative business models emerging in e-health in the future, Pareto-efficiency may be restored.

7.5 Emergent Themes and Patterns

7.5.1 Government Mandate and Incentives to Mitigate Participation and Cooperation Dilemmas

To encourage e-health adoption among healthcare providers, it would help if there is a clear mandate from the government with provision for incentives. Hill and

Powell (2009) insist that a national-level agenda is necessary to make e-health a reality and Gajanayake, Sahama and Iannella (2013) identify government incentives as one of the critical factors for the success of national-level e-health implementations. The two case studies undertaken in this research also serve to prove this point. While the US case study demonstrates that incentives for EHR adoption and its meaningful use may work to some extent, the Singapore case study shows that without incentives the healthcare providers (especially private sector GP clinics) may not be inclined to invest in EHRs. This is further substantiated by an Accenture (2012) report which showed that Singapore lags behind the US in terms of healthcare IT adoption as well as health information exchange both in the primary and secondary care sectors. The gap between the two countries is even more conspicuous in the primary care sector. The lack of integration at the primary care level is also a matter of concern as seen from the Singapore case study, and, as suggested by some informants who participated in this study, incentives from the government might facilitate better integration. It is thus evident that the participation dilemmas observed and discussed in the case studies can, to a great extent, be addressed through a government mandate and some provision for incentives. Singapore may have to consider incentivizing the private sector healthcare providers who make up a substantial 80% of the primary care sector in order to secure their participation in e-health. It may be recalled from the WHO statistics that the Singapore government foots only about 40% of the total healthcare expenditure, which is far lower than the average of 72% as in the case of most developed economies (Britnell, 2015). It may be an indicator that the health policy makers in Singapore should consider further incentivizing the players so as to augment the benefits of its investments in the NEHR, H-Cloud and HealthHub, and, more significantly, to realize the dream of '*One Patient One Record*'.

A key finding corroborated across cases as well as between research and practice is that government incentives can boost e-health adoption among healthcare providers. However, it does not suffice if they simply invest in EHRs all for the sake of realizing some intra-organizational benefits limited to their institutional settings. There will be more value to capture if they co-operate with other healthcare providers to mutually exchange patient data so as to be able to create and maintain a longitudinal patient health record. This would ultimately pave the way for a paradigm shift from provider-centrism to consumer-centrism. This, however, is by no means easy to accomplish as has been observed in the case studies. It may be recalled that competition among healthcare providers often results in cooperation dilemmas and leads to narrow networks, and information asymmetry. It has been suggested in the literature that legislation has the power to maneuver providers towards greater coordination (Fuchs, 2009) and incentives for sharing data which could work as well (Neupert, 2009). Ramesh, Wu & Howlett (2015) acknowledge that designing and enforcing effective healthcare regulations may prove unwieldy, and may necessitate an offer of substantial healthcare subsidies from governments. However, providing incentives to healthcare providers may not be sustainable in the long run. Bandyopadhyay et al. (2012) suggest that the government should intervene in situations where the other stakeholders lack the incentive to act in the best interest of all concerned. Van Limburg et al. (2011) caution that inadequate legislation and a status quo of the traditional roles and dependencies in healthcare will only bolster inefficiencies. Thus, it is clear that a strong government mandate is indispensable to get the players, particularly the healthcare providers, actively involved in the e–health ecosystem. The second-best option may be for a neutral private entity to discover a business model that will incentivize data sharing among the players (Bandyopadhyay et al., 2012). In the context of the US, such an entity is the eHealth Exchange, a public-private initiative with a mission to improve patient care. It is remarkable that, in spite of the highly fragmented nature of the US' healthcare industry, about a third of the country's population is already supported by this platform. In Singapore, this neutral entity is IHis, a wholly owned subsidiary of MoH Holdings with an arm's length relationship with the regulator.

Accenture (2012) suggests three stages in a country's journey towards connected health. Figure 7.4 depicts these stages.

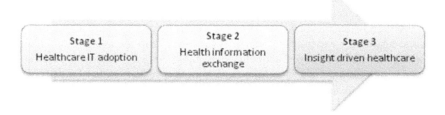

Figure 7.4. Stages of connected health

The first stage involves planning a digital infrastructure, building it and actually using it to create health data in the form of electronic health records for healthcare consumers. The second stage involves exchanging of health information among healthcare providers, as well as between healthcare providers and patients, for effective and efficient patient care. The third stage involves advanced use of health data for purposes such as analytics and informatics that support population health management and clinical decision making.

7.5.2 Population Health Management and Big Data

It may be noted that population health management and 'big data' are two key themes that emerged from the Singapore and the US case studies. Population health management refers to a systematic collection and aggregation of patient data in a form that lends itself to easy access and interrogation for the purpose of sophisticated health data analytics (Accenture, 2012). Such aggregated personal data is so valuable that it is acknowledged to be an asset class to acquire which

Cross-case Analysis and Findings **149**

many organizations including Google and Facebook so keenly compete (Hafen et al., 2014). Population health can be effectively managed by taking advantage of the staggering amount of health data captured by a variety of sources including EHR systems, wearable devices and sensors, which, when aggregated are referred to as 'big data'. The aggregation of these individual health data sets into big data can robustly support the development of effective strategies intended to improve healthcare quality, access and outcomes, bringing about, over a period of time, an overall improvement in population health (Kayyali et al., 2013). Developing the capability to create and use big data to power their population health management strategies is an important goal for Singapore and the US alike.

7.5.3 Preventive Healthcare

Population health management includes among other things, promoting preventive healthcare through patient engagement and empowerment which in turn, can lower healthcare utilization and costs. These are outcomes that are desirable for both countries, and for that matter, for any country in the world. Preventive healthcare is a key area of focus for Singapore which is confronted by the onset of a silver tsunami. If left unchecked, the silver tsunami may place a huge strain on the country's healthcare infrastructure and take it to a point of collapse. According to a recent study by OCBC (Singapore Business Review, 2016), hospital admissions in Singapore are projected to rise to 791,000 yearly, which would imply that public hospitals have a great deal to do in terms of developing their infrastructure so as to be able to cope. Preventive healthcare is particularly crucial for Singapore which advocates personal responsibility for health without which individual healthcare costs may turn out to be phenomenal. It may be pointed out that nearly 55% of Singapore's total healthcare expenditure is out-of-pocket, and the per capita value in Singapore (purchasing power parity adjusted) is at least twice as much as in the US. This state of affairs has even led critics to question if Singapore 'really' offers universal health coverage, as it claims. In the words of Britnell (2015) who has studied healthcare systems around the world, this is the "troubling reality beneath the overall impressive performance" of Singapore's healthcare system (p. 42).

In the case of the US, population health management supported by big data is perceived to have the potential to reduce the country's healthcare expenditure by promoting preventive healthcare, thereby improving healthcare outcomes not only in clinical terms, but also in financial terms. It may be recalled that the ACOs and P-4-P systems recently implemented by the US government's Affordable Care Act are also mechanisms aimed at encouraging preventive healthcare and controlling healthcare costs.

7.5.4 Provider Accountability

Systems such as P-4-P and ACO implemented by the US government make healthcare providers accountable for their patients' health outcomes and compensate them, where necessary, on the basis of how they maximize the use of preventive healthcare to improve health outcomes. Population health management can support these systems by directing physicians towards appropriate evidence-based healthcare protocols (DeVore & Champion, 2011). It is believed that the country's enormous healthcare expenditure can be brought under control through these various complementary measures. The healthcare system costs the country so much that, if it could be put on an island and floated out into the Atlantic Ocean, it would have the fifth largest GDP in the world according to a former adviser to President Barrack Obama (Rosenthal, 2013). Although mechanisms such as P-4-P and ACO are not yet in place in Singapore, trends observed in the course of the case study such as rising healthcare consumerism and quest for transparency in healthcare pricing and outcomes, are predicted to pave the path for accountability as a design characteristic in healthcare systems.

7.5.5 Cloud Technology

Another common theme that emerged from the two case studies is cloud technology. While cloud technology is already being implemented in Singapore, it is expected to take off in the US soon. Singapore has made remarkable strides in the use of cloud technology even to the extent of winning the prestigious DataCloud award for its innovative use of the technology to deliver value (Lee, 2015). Among other benefits that accrue from the use of the technology, Singapore embraced it mostly for its cost-effectiveness and scalability. When a small nation like Singapore can benefit tangibly from the technology, it is conceivable that the US could benefit from the technology a good deal more. The first author recalls being told during the expert interviews that the US has been keenly watching these developments and that it may just be a matter of time before the technology takes off in the country's healthcare industry in a big way. Considering that health data is growing at an exponential rate, the cloud technology may be the way forward for the healthcare industry anywhere in the world.

7.6 Sustainable E-Health Ecosystem: Critical Success Factors

Based on the insights iteratively and incrementally gained through the within-case as well as the cross-case analyses, this monograph addresses RQ2 and recommends that the following critical success factors be taken into account in evolving a patient-centric sustainable e-health ecosystem.:

7.6.1 E-Health Standards for Interoperability

The ecosystem should seamlessly connect all the key players namely, Providers, Payers, Regulators, Vendors, Infomediaries and Citizens. Interoperability is a prerequisite for such an ecosystem design. Considering that healthcare is a public good, the government should lead the development of common technical standards that will facilitate interoperability in the e-health ecosystem. This is made evident by the two case studies. While Singapore has already begun taking steps towards developing a national architecture based on international standards for aggregating health data across diverse clinical systems and data models (Gunapal et al., 2016) the ONC in the US has created a roadmap to achieving nationwide interoperability by 2024 (Krumholz, Terry & Waldstreicher, 2016).

7.6.2 Health Data Custodianship

Since health data is sensitive, the government concerned or its appointed agency should act as the data custodian in order to instill and maintain public confidence. A survey of 20 OECD (Organisation for Economic Co-operation and Development) countries undertaken by Oderkirk, Ronchi & Klazinga (2013) revealed that many of these countries had their national health datasets in the custody of health and/or statistical ministries. It is thus palpable that none other than a government or government-appointed agency can better fulfill the role of such a custodian who will have to carefully and objectively weigh the risk trade-off between individual privacy and public interest in making decisions regarding utilization of health data. In Singapore, this role is fulfilled by HSOR (Gunapal et al., 2016), while in the US, such custodians exist in a complex web at several levels such as states and regions (Oderkirk et al., 2013).

7.6.3 Neutral Infomediary

The infomediary that connects the players in the e-health ecosystem should be a neutral entity that fosters innovation of new business models that incentivize the players for sharing data. It may be a public entity as in the case of Singapore (HealthHub), or a public-private initiative as in the case of the US (eHealth Exchange).

7.6.4 Longitudinal Health Record

It is crucial that the e-health ecosystem is fully integrated, attracting participation and cooperation across the public and the private sector healthcare providers in the primary, secondary, tertiary, intermediate and long-term care sectors. This is to ensure a continuum of care for the patients and foster the creation of a longitudinal health record for them. Moreover, being connected to the whole range of healthcare providers also promotes consumer awareness and choice. It may be recalled that continuum of care was a common theme (concern) for both Singapore and the US.

7.6.5 Accountability

An integrated ecosystem would moreover expose healthcare providers to market forces thus leading to a healthy competition. The competition will manoeuver the providers to assume accountability for their patients' health outcomes and to expect to be compensated on that basis. This will result in greater efficiency in the healthcare market. Given the rising healthcare consumerism and the resultant demand for transparency in healthcare pricing and outcomes evidenced in both case studies, an integrated healthcare system that has the potential to promulgate consumer choices through efficiency, transparency and accountability, is the way forward.

7.6.6 Balancing Efficiency and Fairness

It is known that there is always a tradeoff between efficiency and fairness (Sharma and Bhattacharya, 2013). An efficient e-health ecosystem may not be fair to the healthcare providers who have to invest in technology, share the benefits of their investment with the other players, and moreover, be exposed to competitive forces in the market. Howver recourse to mitigate the 'unfairness' meted out to them by the disruptive innovation of e-health, as has emerged from the case studies and validated literature, may be made available in the form of government incentives (Ramesh et al., 2015), cost-effective technologies like cloud (Nigam & Bhatia, 2016), and innovative business models (Bandyopadhyay et al., 2012) presenting new business opportunities. Disruptive innovation has ushered in affordability and accessibility in many industries ranging from Steel to Finance and it may just be the "right prescription" for the healthcare industry, "a treatment that is desperately needed and long overdue"' (Hwang & Christensen, 2008, p. 1335).

The e-health ecosystem should therefore incorporate mechanisms to foster transparency of information about the players such as healthcare providers, vendors and payers/insurers in its design, which will enable healthcare consumers to evaluate these players based on parameters that may matter to them such as cost, quality and effectiveness, and empower them to make informed decisions regarding their healthcare and healthcare related purchases. This will also improve efficiency in the market.

7.6.7 Citizen-Centrism and Population Health Management

Another trend that emerged from the case studies and is widely supported in literature is that a futuristic e-health ecosystem should be more than patient- or consumer-centric. It should in fact be designed to be citizen-centric so as to offer scope for effective population health management and preventive healthcare. It should encourage citizens to contribute health data, perhaps through an independent PHR interoperable with the EHR systems in the ecosystem.

7.6.8 Big Data and Cloud Technology

It should be part of the e-health ecosystem's design to create and use big data to support development of strategies to improve population health outcomes. While Singapore is already on its way to building a central database that would facilitate big data analytics to support population health management activities (Gunapal et al., 2016), big data is increasingly being recognized as a priority area in the US (Wang & Hajli, 2017). The sustainability of such an ecosystem may be enhanced if it is cloud-based.

7.6.9 Potential Additional Key Players

With the take-off of wearable healthcare devices, mobile health apps and big data, the e-health ecosystem may expand to include new players such as communication network providers and pharmaceutical companies. Hence provisions for these players to be part of the ecosystem should be built into the ecosystem.

7.7 Chapter Summary and Recap

As seen above, this chapter comprises a cross-case analysis of the Singapore and the US case studies undertaken as part of the research and the findings inferred from the analysis. The cross-case analysis was conducted to address RQ2 which sought to identify the critical success factors that must be considered for ensuring sustainability of e-health ecosystems. The chapter began with a comparison of the healthcare systems in both countries using latest statistics on key healthcare indicators from the WHO. Similarities and differences across these two healthcare systems as observed from the within-case analyses in chapters 5 and 6 were highlighted. Common patterns observed across these two diverse cases that were selected for the study on the basis of the purposeful sampling strategy of 'maximum variation' were of particular interest, because such patterns suggest a universal set of characteristics applicable in the context of any healthcare system in the world. Significant themes that emerged in the process were discussed, theoretically integrated, and, finally, critical success factors for developing a sustainable citizen-centric (as opposed to patient- or consumer-centric in the conceptual framework) e-health ecosystem were derived and proposed.

The next chapter presents a summary of conclusions from the study, highlights the practical implications of this study, underscores the study's contribution to the body of knowledge on e-health, acknowledges the study's limitations, and finally, concludes with suggestions for further research.

CHAPTER

8

Discussion and Conclusions

This chapter discusses in brief the inferences and conclusions from this research and thus completes phase 3 of this research. The findings that provide answers to the research questions raised in phase 1 are summarized and the implications of these findings are discussed. The chapter, moreover, highlights the significance and relevance of this study, and its contribution to the body of knowledge on e-health. Finally, the limitations of this study are acknowledged and suggestions for further research are offered.

8.1 Summary of Key Findings: Answers to Research Questions

Research Question 1 (RQ1): *Who are the key players in a patient-centric e-health ecosystem and what are the potential values they can create and capture through digital data flows?*

8.1.1 Key Players

The findings from the case studies confirm five of the six categories of key players identified in the conceptual model, namely Providers, Payers, Vendors, Regulators and Infomediaries. The sixth player, originally identified as Patients / Consumers in the conceptual model, eventually came to be relabeled 'Citizens' upon validation of the conceptual model using the case studies of Singapore's NEHR and the US' HITECH program. It was observed that e-health initiatives in either case were directed towards a more comprehensive group of stakeholders who could be appropriately termed citizens rather than patients or healthcare consumers. The decision to rename the sixth player was also made by way of acknowledging the value that accrues from population health management and preventive healthcare activities which improve health outcomes on the one hand and reduce healthcare costs on the other. In a way, this has caused a paradigm shift - from patient- or consumer centrism to citizen-centrism.

As for the other five categories of players, they are no doubt observable in either case study with the exception of the category Payers in the case of

Discussion and Conclusions   **155**

Singapore. However, these players are not well-connected on a unified platform as envisaged in the conceptual model. This may be attributed to the fact that both the ecosystems are still evolving. Moreover, to forge connections among the key players and foster sharing of health information among these players, the participation and cooperation dilemmas discussed in the conceptual model and observed to be at play in both the cases need to be resolved. For this to happen, recognized interoperability standards must be promoted, adopted and implemented. Also, more importantly, there must be a clear mandate from the government to drive e-health.

8.1.2 Values Created and Captured

In view of the facts mentioned in 8.1.1, it may be observed that the potential for the key players to create and capture the significant values identified in the conceptual model can only be partially realized as at this point of time. It may be recalled that section 4.1.5, 'Value Analysis of the E-Health Ecosystem', presented an analysis of the significant values that can be potentially generated in a well-connected e-health ecosystem. However, the findings from the two case studies reveal that both ecosystems have yet to develop the level of connectedness required for the value flows envisaged in the conceptual model to be realized.

In Singapore, the NEHR connects the public sector healthcare providers and just a section of the primary healthcare providers in the private sector. Given that 80% of primary healthcare is provisioned by private healthcare providers, it needs to be said that the integration of the healthcare system which actually begins at the level of primary care, is at risk. This diminishes the possibility for a longitudinal health record to be created. A longitudinal health record supports a continuum of care by facilitating health data exchange and reuse across the entire spectrum of healthcare providers in both the public and private sectors. Significant values that can be generated in an e-health ecosystem on account of a longitudinal health record are lost as a result. The emerging infomediary namely HealthCloud attempts to bring together various categories of players such as the Patients, Providers, Regulators and Vendors, but how well and when connectivity among all these players will be enabled is still unclear at this point in time. It also remains to be seen if Payers will be included in the ecosystem at a later point of time. More significant values can be created only if all the above discussed player categories are made part of the ecosystem with connectivity forged among them. The ecosystem will then possess the capability to stimulate transparency, competition, cooperation and accountability among the various players, which in turn, will empower consumers by promoting awareness and choices.

In the US, many of the significant values identified in the conceptual model are observed in the ecosystem. However, these values exist at various levels and in a variety of business models, owing to the country's highly fragmented healthcare system that is dominated by private healthcare providers. Many of these values are often restricted to 'narrow networks', which are reluctant to share patient data with other competing networks. This has led to a narrow set of transactions taking place within these networks as against the vast potential of a well-connected e-health ecosystem to support more broad-based and meaningful transactions. As a consequence, the possibility of a longitudinal health record for healthcare consumers, a prerequisite to support continuity of care, is endangered. Attempts are underway to bring all these values onto a unified platform namely the eHealth Exchange, so as to form a nation-wide health information exchange. With about 50% of the US hospitals and more than 100 million patients already on board the platform, the challenge for eHealth Exchange lies in getting the remaining players on board. This can be accomplished only if connectivity among the smaller units of HIEs at the regional / community levels is achieved.

Research Question 2 (RQ2): *What are the critical success factors for developing a sustainable patient-centric e-health ecosystem?*

The cross-case analysis of the two unfolding national-level e-health initiatives in Singapore and the US presented in Chapter Seven, shed light on the critical success factors that must be taken into account for developing a sustainable e-health ecosystem. It facilitates a grasp of the key issues surrounding e-health implementations and an understanding of the ways they can be tackled and resolved, as well.

The two cases studied have significantly focused on the need for e-health initiatives to be undertaken on a national scale and made citizen-centric. A key aspect of such a citizen-centric e-health initiative would be putting in place a longitudinal health record for citizens who are the ultimate beneficiaries of healthcare. A longitudinal health record should among other things be interoperable facilitating ease and convenience of exchange and reuse of health information among the various stakeholders in the process. The case studies show that what is primarily needed to make this happen is a clear vision and mandate from the government concerned. The US government, for instance, adopted a carrot-and-stick approach in respect of EHR adoption and use, which not only facilitated adoption and use of the EHR technology, but also accelerated the process of integrating healthcare services. In the case of Singapore however, for want of a clear mandate and well-defined incentives, EHR take-up among the private sector healthcare providers was much less than expected or hoped for.

It is thus evident that government intervention is crucial for initiating and developing an integrated, citizen-centric e-health. A citizen-centric e-health ecosystem is bound to generate a huge amount of health data and should therefore be designed to support its exponential growth. This can be achieved best by using cloud technology which would likely ensure economy and scalability among other benefits. The huge quantum of data in the system thus designed may then be harnessed for big data analytics which has the potential to drive productivity, innovation, competition, accountability, new values and business models in healthcare.

8.2 Implications for Researchers and Practitioners

The present study casts light on the key issues surrounding the design and implementation of e-health ecosystems, suggesting how these issues may be addressed, and, finally, deriving a critical success factor framework which is crucial for the success of e-health implementations. The larger picture that emerges from such a framework developed on the basis of lessons learnt from the two case studies of national-level e-health implementations will likely benefit both ongoing and prospective initiatives, particularly in developing countries that cannot afford the luxury of failed experiments.

Extrapolating the findings presented in the previous section, it may then be stated that any futuristic e-health ecosystem, needs to be a national-level system so as to be able to cater to its citizens, the ultimate beneficiaries of healthcare.

As a minimum digital platform, the citizen-centric ecosystem should:

(i) provide citizens with ubiquitous access to their longitudinal electronic health records and afford them the ability to co-create their health profiles;
(ii) connect citizens with all the other players that could possibly support them in their healthcare;
(iii) foster an environment that affords easy access to the information citizens need in order to help them arrive at carefully considered decisions regarding their healthcare;
(iv) reward the other players primarily on the basis of the value they create for the citizens;
(v) prevent players from commercially exploiting citizens' health data and ensure that the citizens' trust in the privacy and confidentiality of their health data is on no account compromised; and
(vi) encourage citizens to play a more active role in their healthcare by creating health awareness and promoting preventive healthcare among them.

Such a citizen-centric e-health ecosystem can only be created by a neutral entity having no conflict of interests.

The natural candidate for this role is a country's government. Needless to say, no private entity would come forward to invest heavily in an ecosystem without expecting to capitalize on it and, possibly, without compromising the interests of the citizens who are the sources of health data, a major asset in the ecosystem. Equally important to remember is the fact that the benefits that accrue from an exchange and reuse of health data cannot be harvested in full except through a mandate from the government with or without a provision for incentives. A citizen-centric e-health ecosystem should therefore be government-led and connect the citizens to the other players who support them in their healthcare such as the providers, vendors, and payers/insurers. Given the vast stores of data that will be available in the ecosystem, it may ideally be hosted on a cloud platform. In short, a sustainable e-health ecosystem will have to be citizen-centric, government-led and cloud-based.

8.3 Contributions and Limitations of the Research

The present study highlights facets of e-health that have hitherto not been explored at all or explored only minimally. It may not be an exaggeration to say that the present study has succeeded in highlighting such crucial facets with a thoroughness not seen in earlier studies. Since 2000 when the term e-health was coined, a fairly large number of studies have emerged: some focusing largely on the relevance of health e-commerce for e-health business models (Parente, 2000; Aggrawal & Travers, 2001; Joslyn, 2001; Wen & Tan, 2003); some examining the potential values that HIEs could create in healthcare by functioning as intermediaries (deBrantes, Emery, Overhage, Glaser & Marchibroda, 2007); and some others highlighting the technical design aspects of the EHR (Raghupathi & Kesh, 2009; Paun et al., 2011). No doubt these studies were significant but the fact remains that they all centered around specific facets of e-health, without broaching the significant issue of a business model that could perpetuate significant values in the ecosystem.

On the other hand, there were studies that drew attention to the lack of evidence in favor of EHRs, in addition (DesRoches et al., 2010; Wiedemann, 2012; Kellermann and Jones, 2013), and the lack of clarity in terms of how a health IT platform can be organized to facilitate an exchange of health information among the stakeholders (Hillestad & Keeler, 2014; Bergmo, 2015).

From the above discussion, two clear observations emerge. First, most of the e-health-related studies are narrowly-focused on a select few facets without much reference to the larger picture of an ecosystem. Second, there is still a lack of demonstrable evidence in terms of how e-health should be organized for viability and sustainability.

It is believed that the present study has made a key contribution by filling some of the above-mentioned gaps in e-health literature. The study has provided a deep insight into the broad range of issues involved in e-health adoption by key players. Some of the significant issues explored in this study are the participation and cooperation dilemmas faced by healthcare providers, lack of recognized interoperability standards to support seamless exchange and reuse of health data, absence of a clear mandate from the government and lack of government incentives to spur EHR adoption. The study moreover derived a critical success factor framework which can be utilized to proactively conceptualize the impediments to sustainable e-health. Such a framework that is based on the successes and failures encountered in e-health initiatives in Singapore (a high-performing country in healthcare) and the US (a low-performing country in healthcare) will prove crucial for designing and implementing viable e-health ecosystems. The framework may particularly benefit developing countries which will sooner or later venture into e-health. A further important contribution of this study is the up-to-date, holistic overview it provides of the e-health ecosystems shaping up in Singapore and the US. All this put together may be said to have significantly added to the current body of knowledge.

Discussion and Conclusions **159**

Before concluding this monograph, it is necessary to acknowledge some major limitations of this study. To begin with, the data from the field, though rich, was limited. Despite the effort made to systematically identify and reach out to as many potential informants as possible, the eventual response rate was less than desired. This may, to some extent, be attributed to the potential informants' lack of familiarity with the topic of e-health which, even today, is still evolving. To make up for the shortage of data from the field, a variety of documents had to be relied upon to develop as holistic a view on the topic of interest as possible. It is a matter of doubt, however, whether such a unified view of the topic could have been achieved even if more data through interviews had been made available. It may be pointed out that documents are also accepted as a major source of data in qualitative studies (Charmaz, 2014). As a matter of fact, documents, as sources of data, are perceived to be more objective and, therefore, more reliable than interviews (Charmaz, 2014).

Another limitation of the current study is that the digital flows in the ecosystem were not variously categorized into data, information and knowledge. Raw data, it may be pointed out, is typically captured as patient records either through a medical practitioner or sensors. This data has to be contextualized and, sometimes, analytics helps in transforming it into information. Knowledge would include the social capital of relationships among practitioners as well as among healthcare consumers in the social media. The present research does not adequately distinguish between the quality of such data, information and knowledge, and does not take into account the fact that the upward transformation from data to information to knowledge may bring about higher value activities.

8.4 Suggestions for Further Research

- A suggestion for further work that emerges from the present study is the need for tracking the values created and enhanced as the digital flows transition from data to information to knowledge.
- Equally challenging would be the task to probe how Singapore's e-health ecosystem can help address the issues of high per capita out-of-pocket expenditure and rising hospital readmission rate. Given the country's ageing population and the impending silver tsunami, acquiring the capability to create and use big data to drive population health management initiatives is particularly important for Singapore. The silver tsunami confronting Singapore is predicted to place a huge strain on the country's public healthcare infrastructure which, at its current capacity, can only meet one-fourth of the projected demand for hospital beds by 2030 (YahooNews, 2016). Preventive healthcare, an aspect of population health management, is hence crucial for Singapore, more so, because of the country's emphasis on personal responsibility for health. It may be recalled from the discussion in the previous chapter that the per capita out-of-pocket expenditure in Singapore is twice as much as that in the US and that the country's hospital

readmission rate has steadily risen to a level comparable to that in the US. Hence research is much needed in the areas mentioned above.

- Yet another suggestion worth considering would be to organize a much larger study along similar lines so as to confirm the validity of the critical success factors of a sustainable e-health ecosystem as identified and highlighted in this research. Equally commendable would be an attempt to evolve alternative models of the e-health ecosystem based on big data and analytics.

It is hoped that the present study will help inspire improvements in the present model of e-health for the benefit of all stakeholders involved in the ecosystem. As Sharma & Kshetri (2020) have cautioned – much research remains to be done in the realm of digital healthcare and how it supports the United Nations' Sustainable Development Goals (SDGs) pertaining to healthcare. In the meanwhile, the inferences from this study give one ground to hope that a health cloud is not just a viable proposition, but one capable of transforming the healthcare industry all over the globe, possibly culminating in 'collaboration for citizen-centric value' rather than 'competition for dollars'.

Appendix A

Expert Interview Template

Dear Sir/Madam,

I am a doctoral student from Nanyang Technological University, Singapore. I am working in the area of e-health, and in particular, my research interest is in **e-health business models**. I have done some preliminary work on the e-health ecosystem, based on which I have identified the primary healthcare market players and the values they create[*] and capture[*] in the e-health market space. I am writing to seek help from knowledgeable practitioners like you, to validate my preliminary work which was primarily based on literature and industry observations. Such validation will help me focus my research efforts on the key players in e-health and the really critical issues facing them. I sincerely hope you will be able to assist me in this research.

Please let me have your judgment after carefully going through the background provided below.

I: E-Health ecosystem

The healthcare market has players at two levels – **primary** and **secondary**. At the primary level of the healthcare continuum are market players who use health information to provide patient care directly or indirectly (by supporting direct providers of patient care). These primary healthcare market players have been described in Table A.1.

Secondary market players on the other hand, use health information in roles other than direct and indirect patient care activities. These are public and private organizations focusing on public health, patient autonomy and clinical case management activities (Busch, 2008).

Based on this background, we have attempted to diagrammatically represent the e-health ecosystem which is as shown in Figure A.1.

---

[*] The terms "value created" and "value captured" are defined on page 3, in section **Background II.**

162   *Appendix A*

Table A.1. Primary healthcare market players

Players	Role
Patients	Recipients of health services, often labeled by their financial status: insured or uninsured, privately or publicly insured, with or without financial assets, for the purposes of Financial Case Management[1] and Clinical Case Management[2] (Busch, 2008).
Providers	Any clinical setting and professional staff that designs, implements, and/or executes any healthcare initiative which may be part of a wellness or illness program (Busch, 2008).
Third-party vendors	A large group of diverse market players who play a supporting role to the Providers in their provision of healthcare. These players include health IT vendors, medical equipment vendors, pharmaceutical vendors, transportation services, laboratories, legal systems, billing agents (Busch, 2008).
Payers	Any entity that processes the claims payment transactions of healthcare episodes on behalf of plan sponsors. A plan sponsor is an entity that funds a health program – private insurance plans, government-sponsored plans, employer-sponsored plans (Busch, 2008).
Infomediaries	Organizations that gather pertinent health information from various sources, and syndicates, aggregate and distribute it to foster patient-centric health care (Busch, 2008; Morales-Arroyo & Sharma, 2009).
Regulators	Public and private organizations that develop capabilities for standards-based, secure and confidential exchange of health information to improve the coordination of care among the e-health market players (Blumenthal, 2009).

[1] Discipline of creating a financial plan to meet the patient's healthcare needs (Busch, 2008)

[2] Includes current healthcare initiatives and past treatment regimes (Busch, 2008)

Questions:

In your judgment, would you agree with our identification of the six primary healthcare market players? Please point out any omissions or redundancies and state your reason(s) correspondingly.

Does the e-health ecosystem above adequately model reality?

Would it be a fair assessment to state that e-health would only work if it brings about cost efficiency as well as improved services?

II: Value created vs. value captured by primary healthcare market players

Value is "created" when a player develops its core competencies, capabilities and advantages to perform work activities that differentiate it from competitors,

Appendix A

Figure A.1. E-health ecosystem

and "captured" when the firm derives economic returns in relation to the value it creates (Shafer, Smith & Linder, 2005).

In the context of e-health, every player in e-health "creates a value" that benefits other players, thereby giving rise to a compelling proposition in the e-health market space to collaborate to "create and capture value" rather than "compete for dollars". However the outcomes of such a proposition may vary from player to player, thus begging the question if the "value captured" justifies the "value created", for all the players involved.

For a sustainable e-health market space, it is critical that the participation of each player be justified in "creating" values for the network in exchange for a compelling value that can be "captured" from the network.

Table A.2. Value created vs. value captured

Primary healthcare market players	Value created	Value captured
1. Patients	□ Health data (Neupert, 2009) □ Personal health records (Peters et al., 2009)	□ Ubiquitous access to health records (Burkhard, 2009) □ Patient-centric healthcare (Burkhard, 2009)

		□ Healthcare ecommerce (B2C, C2C) (Parente, 2000; Eysenbach, 2001)	□ Increased healthcare choices (Joslyn, 2001) □ Lower healthcare costs (Wen & Tan, 2003) □ Empowerment to make informed choices (Purcarea, 2009)
2.	Providers	□ Electronic medical/ health records (Hill et al., 2007) □ Improved quality of care (Walker et al., 2005) □ Tele-healthcare delivery (Hill & Powell, 2009) □ Evidence-based medicine (Busch, 2008) □ Preventive healthcare (Chang et al., 2009) □ Medical informatics (Raghupathi & Kesh, 2009) □ Reduced medical errors (Vishwanath & Scamurra, 2007)	□ Improved operational efficiencies (Wen & Tan, 2003) □ Enhanced cost-effectiveness (Vishwanath & Scamurra, 2007) □ Training and education of physicians (Wickramasinghe, Fadlalla, Geisler & Schaffer, 2005) □ Cost-effective e-procurement of medical supplies (Wen & Tan, 2003) □ New business opportunities created by the eHealth network (B2B, B2C) (Parente, 2000)
3.	Third-Party Vendors	□ Direct access to products/services by customers (Wen & Tan, 2003) □ Targeted marketing facilitated by data aggregation (Parente, 2000)	□ Electronic interface to access and target the gamut of healthcare customers - healthcare providers, patients, insurers (Wen & Tan, 2003) □ Decreased marketing costs (Wen & Tan, 2003) □ Reduced R & D costs (Wen & Tan, 2003) □ Increased business opportunities (B2B, B2C etc.) due to direct access to a large pool of healthcare customers (Payton, 2003)
4.	Payers	□ Electronic interface for claims management (Walker et al., 2005) □ Electronic interface for direct purchase of plans by customers bypassing agents (Whitten et al., 2001)	□ Improved efficiency in claims handling (Wen & Tan, 2003) □ Cost savings (Walker et al., 2005) □ Easier implementation of healthcare regulations (Wen & Tan, 2003)

Appendix A **165**

		□ Research and analytics facilitated by data aggregation (Wen & Tan, 2003) □ Targeted information delivery to customers facilitated by data aggregation (Parente, 2000)	□ Increased business opportunities (B2B, B2C etc.) due to direct access to a large pool of healthcare customers (Parente, 2000)
5.	Infomediaries	□ Total digital health systems (Raghupathi & Kesh, 2009) □ A digital platform to facilitate exchange and reuse of health data among players in the healthcare market (Raghupathi & Kesh, 2009; Morales-Arroyo & Sharma, 2009) □ Improved efficiencies for players in the healthcare market (Wen & Tan, 2003) □ Reduced costs for players in the healthcare market (Wen & Tan, 2003)	□ New markets brought about by syndication, aggregation and distribution of health data (Morales-Arroyo & Sharma, 2009) □ New business opportunities and revenues to be tapped in the form of subscription fees, advertisement revenues, e-commerce transactions (Parente, 2000)
6.	Regulators	□ Improvement in quality of healthcare (Blumenthal, 2009) □ Regulations to protect privacy/security of health data (Blumenthal, 2009) □ Standards for interoperable healthcare systems to facilitate exchange and reuse of health data (Goldstein, 2009)	□ Improved health of populations (Blumenthal, 2009) □ Improved efficiency of healthcare systems (Blumenthal, 2009) □ Lower healthcare expenditure (Hill & Powell, 2009)

Questions:

The following questions pertain to Table 2.

Do you think we have adequately identified the sources of values **created** and **captured** by the six primary healthcare market players? If not, please state the player(s) and specify your reason(s).

Do you think the "value created" is *typically greater than* the "value captured" for any of the six players? Please state the player(s) and specify your reason(s).

Do you think the "value created" is *typically less than* the "value captured" for some players? Please state the player(s) and specify your reason(s).

In your judgement, who among the six players *captures the greatest value* from e-health? And who *creates the greatest value*?

In your judgement, who among the six players *captures the least value* from e-health? And who *creates the least value*?

Creating more value than a player captures is not sustainable. Do you agree?

Capturing more value than a player creates may be profitable for a given time, but not sustainable in the long run. Do you agree?

The e-health eco-system can grow only if there is fairness and efficiency. Do you agree?

Thank you for sharing your expert views.

Appendix B

Expert Interviews Coding Scheme

I. E-Health ecosystem

I. Q 1: In your judgement, would you agree with our identification of the six primary healthcare market players? Please point out any omissions or redundancies and state your reason(s) correspondingly.

S/N	Expert response	Code
1	Yes, I do agree with the market players. I guess Patients would be just one of the categories in another broader one i.e recipients. So I think it would be appropriate if you can name the category differently for e.g. Recipients or Customers/Consumers. I suggest this as not all the recipients are actually patients.	AGREE WITH SIX PRIMARY PLAYERS IDENTIFIED HEALTHCARE CONSUMERS MORE INCLUSIVE THAN PATIENTS
2	Not sure if Infomediaries are necessary. In my opinion primary players are the State, Employers and the Society. Employers gain productivity through surveillance.	INFOMEDIARIES NOT A KEY PLAYER.
3	Yes	AGREE WITH SIX PRIMARY PLAYERS IDENTIFIED.
4	Also need to consider research institutions, medical education systems, Universities, scientists – they seem to affect patient care directly with their treatment options.	AGREE WITH SIX PRIMARY PLAYERS IDENTIFIED. RESEARCH INSTITUTES, UNIVERSITIES ALSO KEY PLAYERS
5	The six players identified covers in concept the entities that participate in the primary healthcare market	AGREE WITH SIX PRIMARY PLAYERS IDENTIFIED
6	You may want to include Next-of-Kin (NOK) who accompanied their parents or friends to seek treatment. Especially, those who are	AGREE WITH SIX PRIMARY PLAYERS IDENTIFIED.

	elderly, information (treatment plan) usually flow to NOK rather than the patients. Most of the time, it is the NOK who makes the decision on behalf of the patients.	NEXT-OF-KIN ALSO A KEY PLAYER
7	I would put pharmaceuticals as a separate category as a primary healthcare player, given their direct impact on healthcare in drugs provision and price management.	AGREE WITH SIX PRIMARY PLAYERS IDENTIFIED. PHARMACEUTICALS ALSO A KEY PLAYER.
8	Your primary market is all right (that is you have rightly identified the 'players') but in your secondary market you have identified 'games' rather than the 'players'.	AGREE WITH SIX PRIMARY MARKET PLAYERS IDENTIFIED SECONDARY MARKET PLAYERS ARE JUST FUNCTIONS THAN ROLES.
9	In general Yes; however, overlapping roles are found. Depending on level of detail, providers can also be Infomediaries as can the Government.	AGREE WITH SIX PRIMARY MARKET PLAYERS IDENTIFIED OVERLAPPING ROLES FOUND DEPENDING ON LEVEL OF DETAIL PROVIDERS CAN PLAY ROLE OF INFOMEDIARIES
10	Yes	AGREE WITH SIX PRIMARY MARKET PLAYERS IDENTIFIED.
11	No. There are levels of data use, but the levels can and should be separated along personal/ functional lines: between patient and care providers, patient and specific direct-care "vendors" (this is a vague term). Regulators should NEVER have access to personal health data unless in the aggregate. Vendors generally should not have full access to all health data, but may have access to select relevant-to-activity information. Payers should be separated from health data in ways that promote efficiency of care, not efficiency of payer metrics. So Primary market players should be limited to patients and doctors/care providers, and possibly patient-centric Infomediaries (not industrial or legislative).	DO NOT AGREE WITH SIX PRIMARY MARKET PLAYERS IDENTIFIED PERSONAL HEALTH DATA IS SENSITIVE. ACCESS TO HEALTH DATA MUST BE WELL DEFINED AND CONTROLLED.

Appendix B **169**

	Secondary market players are mostly vendors, regulators, industry-level Infomediaries, and others that you have listed on that level.	
12	"Patients" is a narrow definition. "Provider" should include patients, their family and sometimes their domestic helpers too as they play an active role in their healthcare. Research institutes and universities need a special mention.	THE TERM 'PATIENT' IS NARROW. PATIENTS, THEIR FAMILY, AS WELL AS HELPERS CAN ALSO BE CONSIDERED AS PROVIDERS. RESEARCH INSTITUTES, UNIVERSITIES ARE ALSO KEY PLAYERS.

I. Q 2: Does the e-health ecosystem above adequately model reality?

S/N	Expert response	Code
1	Yes, the model captures almost all the aspects of the ecosystem.	CLOSE TO REALITY
2	To some extent	CLOSE TO REALITY
3	Yes	MODELS REALITY
4	Yes	MODELS REALITY
5	In essence, Yes.	MODELS REALITY
6	There is a trend driving into population health, I don't know how your model will fit it. Patients → Hospital → GP/Polyclinic/Step-community hospital → Homecare Hospitals are decanting stable patients back into community to avoid choking up the hospital resources.	DOES NOT DIRECTLY MODEL SINGAPORE. DOES NOT CONSIDER POPULATION HEALTH.
7	-	NO RESPONSE
8	I feel the model is very USA specific.	CLOSE TO USA.
9	In general Yes; however, overlapping roles are found. Depending on level of detail, providers can also be Infomediaries as can the Government.	MODELS REALITY. OVERLAPPING ROLES FOUND.
10	Yes	MODELS REALITY
11	No. There are two realities: the current model in the US (which is largely driven by industrial interests, not the health of the nation's citizenry) and a more ideal state as modeled by other countries, including Canada. It's important for you to clarify if you're working toward an extension of the current industry-centric model or otherwise.	NEED TO CLARIFY IF THE MODEL IS AN EXTENSION OF CURRENT MODEL IN THE UNITED STATES.

| 12 | "Provider" should capture the whole range including traditional providers e.g. TCM practitioners, acupuncturists. | MODELS REALITY.

INCLUDE INFORMAL PROVIDERS. |

I. Q 3. Would it be a fair assessment to state that e-health would only work if it brings about cost efficiency as well as improved services?

S/N	Expert response	Code
1	I wouldn't fully agree with the statement. E-health is actually pretty broad a word and classifies many more things than only the patient doctor relation which is the main place where the cost comes into the picture. I would see the ecosystem to be more of a platform for the "Continuing Medical Education" concept. It enables the R&D of a lot more than just patient care.	DO NOT FULLY AGREE. E-HEALTH A BROAD TERM. COVERS MORE THAN PATIENT-DOCTOR RELATIONSHIP PLATFORM PLATFORM SHOULD SUPPORT CONTINUING MEDICAL EDUCATION AND R&D
2	No. Does not capture the externalities related to a healthy population. e.g. chronic disease prevention, disease management & control, and wellness.	DO NOT AGREE E-HEALTH SHOULD ADDRESS POPULATION HEALTH ISSUES – WELLNESS, PREVENTION, DISEASE MANAGEMENT
3	Yes	E-HEALTH WORKS WELL WHEN IT IS COST EFFICIENT WITH IMPROVED SERVICES.
4	Absolutely must	E-HEALTH WORKS WELL WHEN IT IS COST EFFICIENT WITH IMPROVED SERVICES.
5	For e-health to work, the various key stakeholders must achieve some forms of wins. For example, patients must benefit from the implementation (e.g. improved level and quality of care), vendors must have sustainable recurring revenue etc. The benefits must be enjoyed across the spectrum or compensated by another entity. The Govt. paying for the infrastructure. Even this must translate to benefits like ability to utilise data for diseases	E-HEALTH WORKS WELL WHEN IT IS COST EFFICIENT WITH IMPROVED SERVICES. MUST BE A WIN-WIN ARRANGEMENT FOR ALL STAKEHOLDERS EXCHANGE AND REUSE OF HEALTH DATA

Appendix B **171**

	trending or sharing of medical records across different providers to achieve a consistent health/medical record for the patient.	
6	Not necessarily	DO NOT FULLY AGREE SERVICE IMPROVEMENT IS CRUCIAL.
7	Definitely a valid assumption, given any business model will work only if there is cost efficiency. It is important to note that cost efficiency does not mean zero cost to the patients. Many models have shown that patients are willing to pay a premium for improved, personalized healthcare services.	E-HEALTH WORKS WELL WHEN IT IS COST EFFICIENT WITH IMPROVED SERVICES. EFFICIENCY DOES NOT MEAN ZERO COST TO PATIENTS. PATIENTS MAY BE WILLING TO PAY A PREMIUM FOR VALUE.
8	Cost will always be an issue as far as healthcare is concerned, and you will never get any buyer if you ask for a trade-off for lower costs but poorer services. So, both need to be looked into, that is improved services at lower costs (higher cost-efficiency). The question now is whether the healthcare system can recover the IT investments that need to be made through highly efficient and improved services.	E-HEALTH WORKS WELL WHEN IT IS COST EFFICIENT WITH IMPROVED SERVICES. ROI FOR PROVIDERS MUST BE ADDRESSED.
9	No. Decision for e-health is a political one. Yes, at the provider level and would add the ability to carry out research.	E-HEALTH IS A POLITICAL DECISION.
10	In order for the e-health to work, it must bring down the cost of health care in remote areas and also improve services. Otherwise, it may not work as people see no value in using e-health. This is the main reason why e-health is not commonly used now.	E-HEALTH WORKS WELL WHEN IT IS COST EFFICIENT WITH IMPROVED SERVICES.
11	No. E-health will be enacted in multiple ways whether it brings efficiencies in costs and services, or not. Right now there are myriad models extending the reach of different power bases without any efficiencies being realized by both sides. In fact, some industry players (insurance companies, for example) are pushing doctors' offices to comply with requirements that are tremendously burdensome on the doctors' offices.	COST EFFICIENCY AND SERVICE IMPROVEMENT DEPENDS ON THE BUSINESS MODEL ROI FOR PROVIDERS MUST BE ADDRESSED.

| 12 | Not necessarily. It should be economically sustainable and add value. E-health would work (equitable). Aggregate benefits to society. So long as the net benefit to society is positive, e-health should work. Provider lose out even in "steady-state" (not just in the initial stage). Govt. should subsidize EHR investment costs. | DO NOT FULLY AGREE E-HEALTH WORKS WHEN THE AGGREGATE BENEFITS TO SOCEITY ARE POSITIVE. ROI FOR PROVIDERS MUST BE ADDRESSED. GOVERNMENT MUST SUBSIDISE INVESTMENT COSTS. |

II. Value Created versus Value Captured by Players

II. Q 1. Do you think we have adequately identified the sources of values created and captured by the six primary healthcare market players? If not, please state the player(s) and specify your reason(s).

S/N	Expert response	Code
1	Yes, the value created and captured seem to be very good. I would still suggest the below:	IDENTIFIED SOURCES OF VALUES CREATED AND CAPTURED ARE ADEQUATE
	(i) For the patient I think it would be good if you can outline the cost and time effectiveness including the fact that the service can be global (Anytime & Anywhere).	INCLUDE COST AND TIME EFFECTIVENESS AND UBIQUITOUS ACCESS AS VALUES CAPTURED FOR PATIENTS
	(ii) For the providers this can be another very good platform for Education.	CAN SERVE AS A PLAFORM TO EDUCATE PROVIDERS
2	Providers – Prevention and wellness	PREVENTION AND WELLNESS VALUES CREATED BY PROVIDERS
3	Yes	IDENTIFIED SOURCES OF VALUES CREATED AND CAPTURED ARE ADEQUATE
4	Yes	IDENTIFIED SOURCES OF VALUES CREATED AND CAPTURED ARE ADEQUATE
5	On the whole, the key values are presented above	IDENTIFIED SOURCES OF VALUES CREATED AND CAPTURED ARE ADEQUATE
6	*Provider:* Value created/Captured: Proven care Community care /Primary care/Population health	IDENTIFIED SOURCES OF VALUES CREATED AND CAPTURED ARE ADEQUATE

Appendix B   **173**

	Value created: Faster care (Reduce appointment wait time if these stable patients can be seen back in the community) Value captured: Able to provide care for much needed/urgent cases	CAPABILITY TO PRIORITIZE ATTENTION TO MORE DESERVING CASES
7	-	NO RESPONSE
8	-	NO RESPONSE
9	Yes, if it was so straight forward, why hasn't EMR actually happened?	IDENTIFIED SOURCES OF VALUES CREATED AND CAPTURED ARE ADEQUATE EMR IS A FAILURE
10	Yes	IDENTIFIED SOURCES OF VALUES CREATED AND CAPTURED ARE ADEQUATE
11	No. E-health is a broad ecosystem where the values created and captured are yet to be discovered. For example, patients can recognize the benefits (value) of having better access to more information about their own diseases, and can (and do) add to community-based research and discovery about possible and alternative treatments. This widens their choice of treatments and providers that's possibly outside the scope of Jocelyn or Purcurea.	SOME VALUES YET TO BE DISCOVERED CO-CREATION OF VALUE BY PATIENTS ALTERNATIVE MEDICINE
12	Yes. Suggest to be more explicit in terms of beneficiaries of value created and captured.	IDENTIFIED SOURCES OF VALUES CREATED AND CAPTURED ARE ADEQUATE IDENTIFY THE BENEFICIARIES

II. Q 2. Do you think the "value created" is *typically greater than* the "value captured" for any of the six players? Please state the player(s) and specify your reason(s).

S/N	Expert response	Code
1	No, it looks pretty fine and well defined.	VALUE CAPTURED CORRESPONDS WITH VALUE CREATED
2	Everyone captures more value than they create	VALUE CAPTURED CORRESPONDS WITH VALUE CREATED
3	None	VALUE CAPTURED CORRESPONDS WITH VALUE CREATED

4	Providers – from adopting an e-health mode of treatment delivery they are yet to reap the full benefits of the high investments. E-health at this point in time is yet to evolve to preserve the productivity of providers.	PROVIDERS CAPTURE LESS VALUE THAN THEY CREATE PROVIDERS INCUR HIGH INVESTMENT COST
5	The value created can only be quantified when it is utilised in the correct context. It is difficult to determine for sure if the value created is greater than value captured without a suitable context provided.	DIFFICULT TO DETERMINE WITHOUT CONTEXT
6	No comments.	NO COMMENTS
7	There needs to be more discussion on the measurement of value created or captured. For example, market supply and demand is a dynamic factor and plays a role in establishing that 'value', besides clinical and economic outcomes.	DIFFICULT TO DETERMINE WITHOUT CONTEXT
8	-	NO RESPONSE
9	-	NO RESPONSE
10	Yes, the value created for the patients will be bigger than value captured as they will then be able to assess to different providers of the health care.	PROVIDERS CAPTURE LESS VALUE THAN THEY CREATE PROVIDERS INCUR HIGH INVESTMENT COST
11	That depends on politics. Right now our system in the US is heavily weighted toward industry benefits and away from patient value and care.	VALUE AND CARE ARE NOT IN FAVOUR OF PATIENTS IN THE US DUE TO POLITICAL FACTORS.
12	Providers create more value than they capture because of EHR investments.	PROVIDERS CAPTURE LESS VALUE THAN THEY CREATE PROVIDERS INCUR HIGH INVESTMENT COST

II. Q 3. Do you think the "value created" is *typically less than* the "value captured" for some players? Please state the player(s) and specify your reason(s).

S/N	Expert response	Code
1	No, Same as before it is well captured.	VALUE CAPTURED CORRESPONDS WITH VALUE CREATED
2	No	VALUE CAPTURED CORRESPONDS WITH VALUE CREATED

Appendix B   **175**

3	None	VALUE CAPTURED CORRESPONDS WITH VALUE CREATED
4	At a certain level, Health IT companies (EHR vendors) have as yet to integrate fully with other legacy or administration systems – compartmentalization within hospital systems exist.	EHR VENDORS ARE CAPITALISING ON LACK OF INTEROPERABILITY ACROSS SYSTEMS WITHIN HOSPITALS
5	The value created can only be quantified when it is utilised in the correct context. It is difficult to determine for sure if the value created is greater than value captured without a suitable context provided.	DIFFICULT TO DETERMINE WITHOUT CONTEXT
6	No comments.	NO COMMENTS
7	-	NO RESPONSE
8	-	NO RESPONSE
9	-	NO RESPONSE
10	-	NO RESPONSE
11	That depends on politics. Right now our system in the US is heavily weighted toward industry benefits and away from patient value and care.	VALUE AND CARE ARE NOT IN FAVOUR OF PATIENTS IN THE US DUE TO POLITICAL FACTORS.
12	Patients are the ultimate beneficiaries of e-health. They create lower value than they capture.	PATIENTS ARE THE ULTIMATE BENEFICIARIES. PATIENTS CREATE LEAST VALUE THAN THEY CAPTURE.

II. Q 4. In your judgement, who among the six players *captures the greatest value* from e-health? And who *creates the greatest value*?

S/N	Expert response	Code
1	I would give this both to the Care Receivers as they are the key players in this ecosystem with the support from the providers. All the other players play the supporting part in the system.	PATIENTS CREATE AND CAPTURE GREATEST VALUE.
2	a. Providers b. Providers	PROVIDERS CREATE AND CAPTURE GREATEST VALUE.
3	As a third-party vendor, I would be tempted to say we create the greatest value. This is because, we are the conduit thru which the services reach the patients in the most effective and efficient manner	THIRD-PARTY VENDORS CREATE THE GREATEST VALUE.

4	Patient captures greatest value – nothing compares to being able to get better health options with e-health – beats hands down any profits that may be earned by any other entity. At this point in time, regulators/Government create maximum value until indices begin to show better health values because of the adoption of e-health.	PATIENTS CAPTURE GREATEST VALUE. REGULATORS CREATE GREATEST VALUE AS THEY INITIATED E-HEALTH ADOPTION.
5	This is really subjective as it depends on different healthcare systems. E.g. If a pharmaceutical company obtains the information and utilizes the data to determine what sort of drugs are commonly used in a certain region, proceeds to produce in bulk to lower costs, since this is beneficial. However, if the pharmaceutical company utilizes the data to create novelty drugs which serve a lesser purpose utilised for revenue generation (e.g. make certain drugs less effective so the dosage will increase etc.), then it is not beneficial.	GREATEST VALUE CREATED OR CAPTURED DEPENDS GREATLY ON THE MOTIVE OF THE PLAYERS.
6	Patient captures greatest value from e-health. Provider creates the greatest value.	PATIENTS CAPTURE GREATEST VALUE. PROVIDERS CREATE GREATEST VALUE.
7	-	NO RESPONSE
8	-	NO RESPONSE
9	Patients and Public	PATIENTS CAPTURE GREATEST VALUE.
10	The patients will capture the greatest values. If the e-health works really well, the government will create the greatest value for them as it will lower the overall cost of the public health care.	PATIENTS CAPTURE GREATEST VALUE. REGULATORS CREATE GREATEST VALUE AS THEY INITIATED E-HEALTH ADOPTION.
11	-	NO RESPONSE
12	a. Patients (followed by Vendors) b. Providers (primary healthcare sector)	PATIENTS CAPTURE GREATEST VALUE. PROVIDERS CREATE GREATEST VALUE.

Appendix B   **177**

II. Q 5. In your judgement, who among the six players *captures the least value* from e-health? And who *creates the least value*?

S/N	Expert response	Code
1	Infomediaries creates the least value and Regulators capture the least value.	REGULATORS CAPTURE LEAST VALUE. INFOMEDIARIES CREATE LEAST VALUE.
2	a. Regulators b. Patients	REGULATORS CAPTURE LEAST VALUE. PATIENTS CREATE LEAST VALUE.
3	The regulators perform several overlapping functions and hence create the least value.	REGULATORS CAPTURE LEAST VALUE. NO RESPONSE
4	Payers create least value – adoption of e-health and other technologies have not reduced cost of healthcare to the patient, premiums are still growing, and benefits shrinking.	NO RESPONSE PAYERS CREATE LEAST VALUE.
5	This is really subjective as it depends on different healthcare systems. E.g. If a pharmaceutical company obtains information and utilizes the data to determine what sort of drugs are commonly used in a certain region, and proceeds to produce in bulk to lower costs, it proves to be beneficial. However, if the pharmaceutical company utilizes the data to create novelty drugs that serve a lesser purpose just to utilize revenue (e.g. make certain drugs less effective so the dosage will increase), then it is not beneficial.	LEAST VALUE CREATED OR CAPTURED DEPENDS GREATLY ON THE MOTIVE OF THE PLAYERS.
6	Payers capture the least value from e-health. Infomediaries create the least value.	PAYERS CAPTURE LEAST VALUE. INFOMEDIARIES CREATE LEAST VALUE.
7	-	NO RESPONSE
8	-	NO RESPONSE
9	Providers create/capture least value in the fee-for-service model.	PROVIDERS CREATE/ CAPTURE LEAST VALUE IN FEE-FOR-SERVICE MODEL.

10	Infomediaries capture the least value from e-health. The patient creates the least value.	INFOMEDIARIES CAPTURE LEAST VALUE. PATIENTS CREATE LEAST VALUE.
11	-	NO RESPONSE
12	a. Providers b. Patients (followed by vendors if unsuccessful)	PROVIDERS CAPTURE LEAST VALUE. PATIENTS CREATE LEAST VALUE.

II. Q 6. Creating more value than a player captures is not sustainable. Do you agree?

S/N	Expert response	Code
1	No, I would say it creates an imbalance but is still sustainable.	CREATING MORE VALUE THAN THAT CAPTURED BY A PLAYER IS SUSTAINABLE DESPITE CREATING AN IMBALANCE
2	Yes	CREATING MORE VALUE THAN THAT CAPTURED BY A PLAYER IS NOT SUSTAINABLE.
3	Yes	CREATING MORE VALUE THAN THAT CAPTURED BY A PLAYER IS NOT SUSTAINABLE.
4	Yes	CREATING MORE VALUE THAN THAT CAPTURED BY A PLAYER IS NOT SUSTAINABLE.
5	It really depends on what value is being created.	SUSTAINABILITY OF E-HEALTH ECOSYSTEM DEPENDS ON WHAT VALUE IS BEING CREATED.
6	Yes	CREATING MORE VALUE THAN THAT CAPTURED BY A PLAYER IS NOT SUSTAINABLE.
7	-	NO RESPONSE
8	-	NO RESPONSE
9	-	NO RESPONSE

Appendix B   **179**

10	Yes	CREATING MORE VALUE THAN THAT CAPTURED BY A PLAYER IS NOT SUSTAINABLE.
11	-	NO RESPONSE
12	Yes	CREATING MORE VALUE THAN THAT CAPTURED BY A PLAYER IS NOT SUSTAINABLE.

II. Q 7. Capturing more value than a player creates may be profitable for a given time, but not sustainable in the long run. Do you agree?

S/N	Expert response	Code
1	Yes, I do agree with this. It again creates an Imbalance but will stumble the ecosystem as the supporting players will find it hard to meet the challenge.	NOT SUSTAINABLE FOR A PLAYER TO CAPTURE MORE VALUE THAN IT CREATES IMBALANCE IN ECOSYSTEM
2	Yes	NOT SUSTAINABLE FOR A PLAYER TO CAPTURE MORE VALUE THAN IT CREATES
3	Yes	NOT SUSTAINABLE FOR A PLAYER TO CAPTURE MORE VALUE THAN IT CREATES
4	Yes, if market forces are allowed to have a free play and regulation (Government) is fair and equitable. Subsidies or cross subsidies can create this imbalance. In healthcare, there are reasons to believe social interests may play an important role from a regulator point of view than profits	NOT SUSTAINABLE FOR A PLAYER TO CAPTURE MORE VALUE THAN IT CREATES
5	It really depends on what value is being created.	SUSTAINABILITY OF E-HEALTH ECOSYSTEM DEPENDS ON WHAT VALUE IS BEING CREATED.
6	May not be applicable in the Singapore context. 70% of patients are subsidized (the medical charges are subsidized by government).	NOT SUSTAINABLE FOR A PLAYER TO CAPTURE MORE VALUE THAN IT CREATES
7	-	NO RESPONSE
8	-	NO RESPONSE
9	-	NO RESPONSE

10	Yes	NOT SUSTAINABLE FOR A PLAYER TO CAPTURE MORE VALUE THAN IT CREATES
11	-	NO RESPONSE
12	No. As long as it is pareto-efficient (i.e. not worse off their status quo)	SUSTAINABLE FOR A PLAYER TO CAPTURE MORE VALUE THAN IT CREATES IF ECOSYSTEM IS PARETO-EFFICIENT.

II. Q 8. The e-health eco-system can grow only if there is fairness and efficiency. Do you agree?

S/N	Expert response	Code
1	Yes, Agree fully.	GROWTH OF E-HEALTH ECO-SYSTEM POSSIBLE ONLY IF IT IS FAIR AND EFFICIENT
2	Yes	GROWTH OF E-HEALTH ECO-SYSTEM POSSIBLE ONLY IF IT IS FAIR AND EFFICIENT
3	Yes	GROWTH OF E-HEALTH ECO-SYSTEM POSSIBLE ONLY IF IT IS FAIR AND EFFICIENT
4	-	NO RESPONSE
5	The e-health eco-system will grow as long as the payer sees justification, the technology is suitable (need not be cutting edge) and the end-users are willing to adopt it.	GROWTH OF E-HEALTH ECO-SYSTEM POSSIBLE ONLY IF IT IS FAIR AND EFFICIENT
6	All your players need to be open and willing to share information. There are too many silos within the healthcare system (including hospitals).	SILOS IN HOSPITALS/ HEALTH CARE SYSTEMS DETERIORATES GROWTH OF E-HEALTH ECO-SYSTEM.
7	Economic sustainability is critical to the growth of an ecosystem. E-health is still in early stage, and new business models are still being developed. At the same time, a concerted effort by the government to establish basic infrastructure, standards and regulation will be necessary.	GOVERNMENT MUST SUBSIDISE INITIAL INVESTMENT.
8	-	NO RESPONSE
9	Suggest you link e-health advances to reimbursement system	GROWTH OF E-HEALTH ECO-SYSTEM POSSIBLE ONLY IF IT IS FAIR AND EFFICIENT

Appendix B **181**

10	Yes, and the government plays a key role in making it work.	GROWTH OF E-HEALTH ECO-SYSTEM POSSIBLE ONLY IF IT IS FAIR AND EFFICIENT
11	Again that depends on who is defining "fair" and "efficient."	PERSPECTIVE OF FAIRNESS AND EFFICIENCY VARIES FOR PLAYERS
12	Yes. Provided there is not a wide disparity in terms of values created and captured	GROWTH OF E-HEALTH ECO-SYSTEM POSSIBLE ONLY IF IT IS FAIR AND EFFICIENT

Appendix C

Case Study Interview Template

Date:

Name:

Company:

Designation:

Dear Sir/Madam,

I am a doctoral student from Nanyang Technological University, Singapore. I am working in the area of e-health, and in particular, my research interest is in determining the design characteristics of a successful **e-health business model**. I sincerely hope you will be able to assist me in this research.

1. What is the status of e-health in Singapore?

2. How would you describe your participation in/contribution to e-health in Singapore?

3. What are the challenges in implementing e-health in Singapore, in your opinion?

4. Who are the key stakeholders/players who should participate in e-health to harness the full benefits of e-health?

5. How can we encourage or ensure participation from these players?

6. How will e-health benefit its stakeholders and most importantly, Singapore?

Thank you!

Appendix D

Case Study Interviews Coding Scheme

Singapore

1. What is the status of e-health in Singapore?

Expert response #1	Code
Singapore is faced with a 'silver tsunami' which is due to impact the cities and states in the next couple of decades. This tsunami will result in severe bed/resource shortages in hospitals, affecting productivity. Considering that 80% of healthcare costs are incurred in the last few years of an individual's life time, healthcare's focus in Singapore has to shift towards providing better intermediate and long term care for the elderly.	SILVER TSUNAMI WILL AFFECT HEALTHCARE PROUCTIVITY. 80% OF AN INDIVIDUAL'S HEALTHCARE COSTS IS INCURRED IN THE LAST FEW YEARS OF THEIR LIFE TIME.
It is also very important to step up 'preventive healthcare' in preparation for an ageing population. This might mean ushering in 'Smart Homes' equipped with monitoring systems to enable management of chronic health conditions. Such 'Smart Homes' may be supported by Cloud, SaaS, and mobile technologies. Another initiative towards this end is N-HELP, which addresses the operational, clinical, and financial needs of participating nursing homes, primarily voluntary healthcare organizations, which helps boost productivity in these outfits. N-HELP is connected to the NEHR (National Electronic Health Record), the patient-centric longitudinal health record envisioned for Singapore, and thereby supports timely data flows of residents' health information across different healthcare institutions. E-health in Singapore is primarily aimed at improving patient experience, increasing productivity in healthcare institutions, and	BETTER INTERMEDIATE AND LONG-TERM CARE FOR ELDERLY. PREVENTIVE HEALTHCARE PREPARES FOR AGEING POPULATION SMART HOMES AID DISEASE MANAGEMENT N-HELP AIMS AT NURSING HOMES' PRODUCTIVITY N-HELP IS LINKED TO NEHR NEHR CAPTURES LATEST HEALTH RECORDS OF RESIDENTS.

promoting interoperability among disparate health information systems. However, Singapore is currently lagging behind countries like USA and Japan in terms of providing Continuum of Care. The model that will work going forward is the health cloud model which will offer Software as a Service (SaaS), and include care coordinators from the various healthcare institutions.	IMPROVED PATIENT EXPERIENCE NO CONTINUUM OF CARE HEALTH CLOUD IS THE FUTURE
Expert response #2	**Code**
Singapore is making good progress in terms of e-health. To realize Singapore's vision of 'One Patient – One Medical Record', the National Electronic Health Record (NEHR) initiative was launched in 2010, and is now in its second phase of implementation. The grand plan is for the NEHR to be eventually migrated to H-Cloud, the Consolidated Healthcare Cloud that will host mission critical systems for all public hospitals, specialty centres and polyclinics. Even primary care providers can connect to NEHR through NEHR web services/CLEO (Clinic Electronic Medical Records and Operations). Although the current focus is on the public healthcare sector, it is expected that the private healthcare institutions will also be consolidated under H-Cloud in due course.	NEHR ENVISIONS ONE PATIENT-ONE MEDICAL RECORD NEHR WILL BE HOSTED ON H-CLOUD PRIMARY CARE PROVIDERS ACCESS NEHR VIA WEBSERVICES PRIVATE HEALTHCARE INSTITUTIONS MAY BE ALLOWED TO SHARE H-CLOUD
The H-Cloud is owned by IHiS which is a 100% subsidiary of MOH Holdings Pte. Ltd., and serves as their project management wing for healthcare projects.	SILVER TSUNAMI SHIFTS FOCUS TO LONG-TERM CARE
The 'silver tsunami' phenomenon about to impact Singapore soon is shifting the focus of healthcare to long-term care. Some initiatives towards this end are Smart Hospitals, Smart Mat, and others. To elaborate on Smart Mat, it is a non-intrusive device that can capture breathing/heart rates, which are important parameters in clinical monitoring. In addition, it can also determine the quality of sleep, and send timely alerts e.g. fall alerts. It is flexible enough to be mounted on mattresses or cushions in home/clinical settings. This may be an answer to the national bed crunch faced in Singapore hospitals which can stretch the waiting time for a hospital bed to up to 12 hours. Proof-of-concept for Smart Mat is underway in the Changi Prison Medical Centre.	SMART MAT IS A NON-INTRUSIVE DEVICE SMART MAT WORKS IN HOME/CLINICAL SETTINGS SMART MAT AIMS TO TACKLE NATIONAL BED CRUNCH SMART MAT IS IN POC STAGE AT CHANGI PRISON MEDICAL CENTRE

Appendix D   **185**

2. How would you describe your participation in/contribution to e-health in Singapore?

Expert response #1	Code
My company, Napier Healthcare is a global company headquartered in Singapore. We are a healthcare 2020 technology enabler focused on patient-centric care delivery in tertiary hospitals, intermediate to long term care institutions, primary care institutions and home care. Our offerings include Hospital Information System (HIS), Loop Patient Referral Management System, Electronic Medical Record (EMR), and solutions for intermediate to long term care. A special mention goes to our Loop Patient Referral Management System which is a cloud solution that helps healthcare providers effectively manage care transitions and patient referral processes.	NAPIER HEALTHCARE IS A TECHNOLOGY ENABLER. PATIENT-CENTRIC CARE OFFERS EMR, HOSPITAL INFORMATION SYSTEM, AND LOOP PATIENT REFERRAL MANAGEMENT SYSTEM LPRMS IS A CLOUD SOLUTION COVERS INTERMEDAITE TO LONG TERM CARE SOLUTIONS

Expert response #2	Code
H-Cloud is owned by IHiS, and facilitated by ST Electronics (Info Software) Systems. Our target is to make the mission critical systems in the various healthcare institutions interoperable by year 2017, and have these migrated to H-Cloud. We have already migrated over 500 servers and over 200 applications of various healthcare institutions, to H-Cloud.	ST ELECTRONICS (INFO-SOFTWARE SYSTEMS) FACILITATES H-CLOUD OWNED BY IHIS. H-CLOUD PROMOTES HEALTHCARE SYSTEMS' INTEROPERABILITY.

3. What are the challenges in implementing e-health in Singapore, in your opinion?

Expert response #1	Code
The current e-health business model may not be exactly sustainable for a number of reasons: In Singapore, the model is pay-for-service rather than pay-for-performance which is the model in the US. In a pay-for-performance model, hospitals that demonstrate high quality of care, will be incentivized through reimbursements. The current model in Singapore does not seem to support care transition and the resulting continuum of care as there is no way for a transparent exchange of information. For example, there is no 'loop' to track general practitioners' (primary care providers') patient referrals to hospitals. *(This is something Napier Healthcare is hoping to address through our Loop Patient Referral Management System.)*	CURRENT E-HEALTH BUSINESS MODEL IS NOT SUSTAINABLE. SINGAPORE MUST ADOPT PAY-FOR-PERFORMANCE MODEL LACK OF TRANSPARENCY IN EXCHANGE OF INFORMATION. CARE TRANSITION AND CONTINUUM OF CARE NEED TRANSPARENCY.

Although the health cloud (H-Cloud) is underway, data residency and sovereignty issues (simply put, 'who owns what?') need to be addressed and resolved to the satisfaction of the stakeholders.	PRIMARY CARE PROVIDERS' PATIENT REFERRALS MUST BE TRACKED. H-CLOUD MAY FACE DATA RESIDENCY, SOVEREIGNITY ISSUES.

Expert response #2	Code
Setting up and maintaining an H-Cloud is a challenge in itself. The various healthcare institutions in Singapore have been functioning like silos, having their own infrastructures, and running their own applications. These silos need to be made interoperable, consolidated, and migrated to H-Cloud, which steps involve heavy initial investments. Although the Government subsidizes these costs to some extent, the healthcare institutions also have to deploy their own funds for the purpose. Although there was some initial resistance to this initiative, it was overcome gradually when these healthcare institutions started seeing the bigger picture and were convinced by the long-term benefits of such a direction. Loss of 'business' to other healthcare institutions as a result of the H-Cloud is not a concern among the public sector healthcare institutions which are already facing a resource crunch and struggling to meet the healthcare needs of the public.	H-CLOUD OFFERS INTEROPERABLE PLATFORM FOR HEALTHCARE INSTITUTIONS. HEALTHCARE INSTITUTIONS SUFFER FROM HEAVY INITIAL INVESTMENT COSTS DESPITE GOVERNMENT SUBSIDY. H-CLOUD OFFERS LONG-TERM BENEFITS. LOSS OF BUSINESS IS NOT A CONCERN TO PUBLIC SECTOR PLAYERS AS THEY WILL BE RELIEVED OF RESOURCE CRUNCH.

4. Who are the key stakeholders/players who should participate in e-health to harness the full benefits of e-health?

Expert response #1	Code
The key stakeholders in e-health are General Practitioners, Vendors, Government, Payers, Patients, and of course, a Service Provider (Infomediaries) who will be able to bring all these stakeholders together on a common platform (an IT system) which will be based on a sustainable business model.	E-HEALTH KEY STAKEHOLDERS INCLUDE GENERAL PRACTITIONERS, VENDORS, GOVERNMENT, PAYERS, PATIENTS AND INFOMEDIARIES. INFOMEDIARIES BRING E-HEALTH STAKEHOLDERS TOGETHER. E-HEALTH NEEDS A COMMON IT PLATFORM TO WORK. E-HEALTH DEPENDS ON A SUSTAINABLE BUSINESS MODEL.

Appendix D   **187**

Expert response #2	Code
The Government, healthcare institutions, insurance companies and the public are the key stakeholders. The NEHR comprises EHR (Electronic Health Record) as well as PHR (Personal Health Record). The health data acquired by healthcare institutions is housed in the EHR. On the other hand, PHR may collect health data in the form of readings from Smart Homes. This can be viewed as health data contribution by the public to the NEHR. Data is also contributed by insurance companies to the NEHR in the form of claims data. All these data sources add to the comprehensiveness of an individual's health record.	E-HEALTH KEY STAKEHOLDERS INCLUDE GOVERNMENT, HEALTHCARE INSTITUTIONS, INSURANCE COMPANIES AND THE PUBLIC. NEHR INCLUDES EHR AND PHR. EHR IS HEALTH DATA ACQUIRED BY HEALTHCARE INSTITUTIONS. SMART HOME INITIATIVES CAN CREATE DATA FOR PHR, WHICH CAN LINK TO NEHR NEHR IS A COLLECTION OF CITIZENS' COMPREHENSIVE HEALTH RECORDS.

5. How can we encourage or ensure participation from these players?

Expert response #1	Code
An interoperable system based on the Continuum of Care Document Architecture (CCDA) devoid of any incentives to stakeholders will not work. The system should provide for some incentives for the key players to cooperate and share information, not necessarily in monetary terms. As for encouraging participation from General Practitioners, a more effective method to get them to leverage the NEHR may be to move	INCENTIVES WILL ENCOURAGE KEY PLAYERS COOPERATION AND INFORMATION SHARING GENERAL PRACTITIONERS MAY BE MOTIVATED TO USE NEHR USING A PAY-PER-USE MODEL OVER MONTHLY SUBSCRIPTIONS.
from the monthly subscription model to a pay-per-use model. A two-way model may be better where a healthcare institution that pays for access to health information in the NEHR, also gets paid/incentivized in some way when it contributes information to the NEHR. This may be a more sustainable model and will also resolve issues related to data residency and sovereignty. Healthcare institutions may also consider charging patients nominally (like $1 per healthcare episode) for creating/maintaining their records. To gain the participation of the general public/ patients, PHR systems may be sponsored by insurers. Insurers may go a step further and incentivize their policy-holders to use the PHR system which will in turn facilitate preventive	HEALTHCARE INSTITUTIONS MAY BE MOTIVATED TO USE NEHR FOR A TWO-WAY MODEL. TWO-WAY MODEL ENSURES CONTRIBUTION TO NEHR AND INCENTVES TO HEALTHCARE INSTITUTIONS. TWO-WAY MODEL PROMOTES SUSTAINABLE E-HEALTH. NOMINAL FEE TO MAINTAIN HEALTHCARE RECORD MAY BE CHARGED TO PATIENTS. INSURERS MAY SPONSOR PHR SYSTEMS PHR SYSTEMS PROMOTE PREVENTIVE HEALTHCARE

healthcare versus the less than ideal diagnostic healthcare which increases costs for the insurers.	INSURERS FACE HIGHER COSTS DUE TO DIAGNOSTIC HEALTHCARE
	PHR SYSTEMS ENCOURAGE PATIENT/PUBLIC PARTICIPATION.

Expert response #2	**Code**
It is our recommendation that the Next Generation Nationwide Broadband Network (NGNBN) which facilitates Smart Homes and Smart Mat, be fully sponsored by the Singapore Government for the benefit of its citizens. 80% of the NGNBN infrastructure is already in place, and the network is expected to be fully operational by 2025.	NGNBN FACILITATES SMART HOME INITIATIVES
	NGNBN MUST BE FULLY SPONSORED BY GOVERNMENT
Call Centres should be established, ideally within hospitals, to monitor smart homes so as to be able to direct healthcare resources to where they are most needed.	HOSPITALS MUST MONITOR SMART HOMES IN ORDER TO MANAGE HEALTHCARE RESOURCES
Homes may be made 'smart' through NGNBN, smart phones and a base kit (like the one for MIO TV) which will enable the public to subscribe to healthcare services they require, and pay for these based on their level of consumption of such services ('pay as you use').	SMART HOME MAY REQUIRE NGNBN, SMART PHONES, AND A BASE KIT.
	PUBLIC MAY WELCOME PAY-AS-YOU-USE MODEL FOR SMART HOME HEALTHCARE SERVICES.

6. How will e-health benefit its various stakeholders and most importantly, Singapore?

Expert response #1	**Code**
E-health will help the Government in keeping healthcare costs low for their citizens. Insurers and Payers will also benefit from the renewed focus on 'preventive healthcare'.	LOWER HEALTHCARE COSTS.
	PREVENTIVE HEALTHCARE IS BENEFICIAL.

Expert Response #2	**Code**
Singapore benefits from e-health if its people benefit from it. E-health will help ease bed/resource crunch in hospitals, pave the way for smarter healthcare – resulting in a shift from reactive to proactive healthcare.	E-HEALTH REDUCES BED CRUNCH.
	E-HEALTH LEADS TO SMARTER HEALTHCARE AND SHIFTS REACTIVE HEALTHCARE TO PREVENTIVE HEALTHCARE.

Appendix D   **189**

The United States of America

1. What is the status of e-health in the US?

Expert response #1	Code
E-health from a medical perspective is fairly advanced since gaining momentum a few years ago, not in the least due to reforms pursued by the Federal Government and cost pressures. Health plan exchanges, mandates for e-health records and insurance cover expansion have spurred technology deployments quadrupling e-health platform deployments since 2011. Government reports that there is almost a 60% compliance with e-health mandates at the physician level and perhaps higher at the hospital level. All this has also led to the adoption of intelligent medical devices that have a higher degree of interoperability and ability to address consumer health informatics. On the ground, we see many physicians still handicapped by lack of proper tools for effective caregiving. There are groups that are resistant to technology intrusion in to their treatment rooms. It is a highly evolving area that has achieved much and still has to travel far to reach a true e-health environment.	FAIRLY ADVANCED FROM A MEDICAL PERSPECTIVE E-HEALTH PLATFORM DEPLOYMENTS QUADRUPLED SINCE 2011 DUE TO FEDERAL GOVT. MANDATES 60% COMPLIANCE WITH E-HEALTH MANDATES AT PHYSICIAN LEVEL ADOPTION OF INTELLIGENT INTEROPERABLE MEDICAL DEVICES TO SUPPORT CONSUMER HEALTH INFORMATCS GROUPS RESISTANT TO TECHNOLOGY LACK OF TOOLS FOR EFFECTIVE CAREGIVING

Expert response #2	Code
80%-90% of the US hospitals have some form of e-health. Individual clinics and community providers collaborate with the nearest hospital to be part of an e-health system. Healthcare providers are in a way compelled to opt for e-health as the US government has mandated this and set a deadline for year 2020. Healthcare providers who do not comply with this mandate will be penalized in the form of cut-backs on reimbursements. On the other hand, healthcare providers who transition to e-health and furthermore meet the criteria for meaningful use, will be incentivized.	80% - 90% OF HOSPITALS HAVE SOME FORM OF E-HEALTH CLINICS AND COMMUNITY PROVIDERS COLLABORATE WITH THE NEAREST HOSPITAL TO BE PART OF AN E-HEALTH SYSTEM HEALTHCARE PROVIDERS ARE FORCED TO OPT FOR E-HEALTH DUE TO FEDERAL GOVT. MANDATES DEADLINE SET FOR 2020 PENALTIES FOR NON-COMPLIANCE THROUGH CUT BACKS INCENTIVES FOR ADOPTION AND MEANINGFUL USE

2. How would you describe your participation in/contribution to e-health in the US?

Expert response #1	Code
1. The biggest challenge was use of paper and entering the same data more than once. We develop interfaces with and between hospital systems and ancillary providers to eliminate this malaise. In addition, we help physicians with no access to interoperable systems by creating work flow platforms that can carry instructions and medical records. Though not the most efficient option, this is a cost-effective option. 2. Secondly, treatment captured in EMR has dragged physician productivity down. We use scribe services to capture the physician notes in the EMR and generate the report for his sign off. Again, not the most efficient, but it allows groups to be compliant with electronic record capture whilst minimizing productivity losses.	BIGGEST CHALLENGE TO SURMOUNT WAS PAPER RECORDS AND MULTIPLE DATA ENTRIES DEVELOP INTERFACE WITH HOSPITAL SYSTEMS AND BETWEEN HOSPITAL SYSTEMS CREATE WORK FLOW PLATFORMS THAT CARRY INSTRUCTIONS AND MEDICAL RECORDS FOR PHYSICIANS WITH NO ACCESS TO INTEROPERABLE SYSTEMS. NOT EFFICIENT BUT COST-EFFECTIVE OFFER SCRIBE SERVICES TO CAPTURE PHYSICIAN NOTES IN THE EMR AND GENERATE REPORTS FOR PHYSICIAN SIGN OFF. NOT EFFICIENT BUT SATISFIES COMPLIANCE
3. EMR technologies are still evolving and hence the CPT/ICD codes that are auto generated by the EMR is subject to review. We resolve this by external manual audit to ensure the EMR can auto correct through a self-learning mechanism. We have alternately developed tools based on Snomed ??CT to auto code based on NLP techniques. Use of our tools can significantly eliminate the coding conundrum for large hospital-based groups.	EMR TECHNOLOGY STILL EVOLVING SO AUTO-GENERATED AUDIT CPT/ICD CODES SUBJECT TO REVIEW RESOLVE THROUGH EXTERNAL MANUAL AUDIT TO ENSURE EMR CAN AUTO-CORRECT THROUGH SELF-LEARNING MECHANISM OUR TOOLS ELIMINATE THE CODING CONUNDRUM FOR LARGE HOSPITAL-BASED GROUPS

Expert response #2	Code
Sutter Health is a not-for profit healthcare organization having a strong presence in Northern California, Oregon, and Hawaii. It is one of the biggest healthcare systems in the US which has more than a million patients covered by its EHR systems and generates 9 billion dollars in revenues. Earnings are re-invested in patient care to achieve quality and innovation in healthcare. The Sutter Health network not only includes	A NOT-FOR-PROFIT HEALTHCARE ORGANIZATION WITH PRESENCE IN N.CALIFORNIA, OREGON & HAWAII ONE OF THE BIGGEST HEALTHCARE SYSTEMS IN THE US MORE THAN 1 MILLION PATIENTS COVERED BY EHR SYSTEMS

Appendix D

hospitals but also individual providers like clinics as well as community providers. It allows the healthcare providers in its network to document their patients' medical information online without having to do any paper work. The individual providers that are part of the Sutter Health network gain access to its EHR system (EPIC) and the resources supporting this system, by paying a subscription fee. Currently, the incentives offered by the US government help these individual providers cover these payments. Earlier, not only the individual providers, but also their patients, had to pay to access their EHR,and message their doctors. However these services are no longer paid services for the patients, the costs being absorbed by Sutter Health. 70-80% of the EHR market in the US belongs to EPIC, AllScripts and Cerner. Some e-health systems other than Sutter Health that are supported by EPIC are Kaiser Permanente, Stanford, UC Davis and John Muir. Since EPIC facilitates easier transfer of medical records from other systems powered by its open source configuration, it is likely to make inroads into several other hospitals in the future. One notable hospital making a switch from their existing system (Cerner) to EPIC is Mayo Clinic. Although such a switch involves exorbitant costs, it is considered a necessary move for the purpose of gaining access to a bigger share of the market. Sutter Health itself uses 24 modules of EPIC and hence receives continuous support from EPIC. While Sutter Health is a non-profit organization, its insurance arm, Sutter Health insurance is a for-profit organization that offers two health plans: HMO that covers only network providers and PCO that covers all providers.

GENERATES 9 BILLION DOLLARS IN REVENUES. EARNINGS REINVESTED TO ACHIEVE QUALITY AND INNOVATION IN HEALTHCARE

NETWORK INCLUDES CLINICS AND COMMUNITY PROVIDERS WHO HAVE ACCESS TO SUTTER HEALTH'S EHR SYSTEM (EPIC) FOR A SUBSCRIPTION FEE

CURRENTLY INCENTIVES OFFERED BY THE GOVT. COVERS THE SUBSCRPTION FEES

EARLIER, THE INDIVIDUAL PROVIDERS AND THEIR PATIENTS HAD TO PAY FOR ACCESS TO THE EHR

PATIENTS NO LONGER HAVE TO PAY, COSTS ABSORBED BY SUTTER HEALTH

70%-80% OF THE EHR MARKET IN US BELONGS TO EPIC, ALLSCRIPTS AND CERNER PUT TOGETHER

EPIC ALSO SUPPORTS KAISER PERMANENTE, STANDFORD, UC DAVIS AND JOHN MUIR

EPIC POWERED BY OPEN-SOURCE CONFIGURATION CAN FACILITATE EASIER TRANSFER OF MEDICAL RECORDS FROM OTHER SYSTEMS

EPIC IS LIKELY TO MAKE INROADS INTO SEVERAL OTHER HOSPITALS IN FUTURE. MAYO CLINIC MAKING A SWITCH FROM CERNER TO EPIC

ALTHOUGH A COSTLY DECISION, NECESSARY FOR GAINING ACCESS TO A BIGGER SHARE OF MARKET

SUTTER HEALTH USES 24 MODULES OF EPIC AND RECEIVES CONTINUOUS SUPPORT FROM EPIC

192   *Appendix D*

| | SUTTER HEALTH IS NOT-FOR-PROFIT BUT ITS INSURANCE ARM IS FOR-PROFIT. |
| | SUTTER HEALTH INSURANCE OFFERS 2 HEALTH PLANS: HMO COVERING ONLY NETWORK PROVIDERS AND PCO COVERING ALL PROVIDERS |

3. What are the challenges in implementing e-health in the US, in your opinion?

Expert response #1	Code
A. Physician acceptance. The aging caregiver population is slow to adapt and adopt latest technologies. Since the technologies themselves are evolving, a large group of physicians/ hospitals are waiting to catch the maturity curve much later.	OLDER PHYSICIANS SLOW TO ADAPT AND ADOPT LATEST TECHNOLOGY
	SINCE TECHNOLOGY ITSELF EVOLVING, PHYSICIANS/HOSPITALS WAITING TO CATCH MATURITY CURVE MUCH LATER
B. Costs. Many small and rural groups do not have the resources to invest and maintain e-health platforms. Though Federal budget support is available, these groups fear long term costs that may not be subsidized.	SMALL/RURAL GROUPS DO NOT HAVE THE RESOURCES TO INVEST AND MAINTAIN E-HEALTH PLATFORMS
C. Delayed Implementation: Many implementations of e-health initiatives fail and result in cost escalation due to improper specifications, poor understanding of the physician business by the e-health company and interoperability with legacy and other systems	THOUGH FEDERAL BUDGET SUPPORT AVAILABLE, SOME GROUPS FEAR LONG-TERM COSTS THAT MAY NOT BE SUBSIDIZED
	MANY E-HEALTH IMPLEMENTATIONS FAIL AND FURTHER ESCALATE COSTS DUE TO IMPROPER SPECS, POOR UNDERSTANDING OF PHYSICIAN BUSINESS AND INTEROPERABILITY WITH LEGACY & OTHER SYSTEMS

Expert response #2	Code
The transition from paper to digital records was achieved by manual data entry as well as scanning of medical documents which were subsequently verified and filed. This was time and resource-intensive. It had to be done this way because labs have a language of their own. The EHR (in this case,	TRANSITION FROM PAPER TO DIGITAL RECORDS ACHIEVED BY MANUAL DATA ENTRY, SCANNING MEDICAL DOCUMENTS WHICH WERE SUBSEQUENTLY VERIFIED AND FILED
	TRANSITION WAS TIME AND RESOURCE INTENSIVE BUT

Appendix D   193

EPIC) understands HL7 and hence is able to "translate" and capture any communication between points A and B. Another challenge is that there is no real-time exchange of information between insurance companies and healthcare organizations. Yet another challenge was in terms of how e-health was received by the physicians. The older generation of physicians were not as receptive to e-health as their younger counterparts.	NO CHOICE AS LABS HAVE A LANGUAGE OF THEIR OWN
	EHR (EPIC) UNDERSTANDS HL7 AND HENCE ABLE TO TRANSLATE AND CAPTURE BETWEEN POINTS A AND B
	ANOTHER CHALLENGE – NO REAL TIME EXCHANGE OF INFORMATION BETWEEN PROVIDERS AND PATIENTS
EHR interfaces had to be customized for the healthcare providers and this was supplemented with user trainings and manuals. If implementing EHRs is a costly affair, maintaining these systems is equally expensive as substantial resources need to be deployed for the purpose.	OLDER GENERATION PHYSICIANS NOT AS RECEPTIVE TO E-HEALTH AS COMPARED TO THEIR YOUNGER COUNTERPARTS
	EHR INTERFACES HAD TO BE CUSTOMIZED FOR PROVIDERS AND SUPPLEMENTED WITH USER TRAININGS AND MANUALS
	IMPLEMENTING EHRS IS COSTLY, MAINTENANCE EQUALLY COSTLY AS SUBSTANTIAL RESOURCES INVOLVED

4. Who are the key stakeholders/players who should participate in e-health to harness the full benefits of e-health?

Expert response #1	Code
a. Physicians groups, Office staff and Hospital Information Management Departments – understand the benefits of a connected healthcare ecosystem for care enhancement, cost reduction in the long run, reduced utilization and better insight into diseases and cure b. Patients: by insisting on e-health initiatives by his/her care giver c. Pharma/clinical R&D : better data availability means analytics and informatics opportunities for developing cure and arresting outbreaks. d. Government: e-health can drive health policies, early warning mechanisms and advisory mechanisms	PHYSICIANS, HOSPITALS UNDERSTAND BENEFITS OF A CONNECTED HEALTHCARE ECOSYSTEM FOR CARE ENHANCEMENT, COST REDUCTION, REDUCED UTILIZATION, BETTER INSIGHTS INTO DISEASES AND CURES PATIENTS SHOULD INSIST ON E-HEALTH INITIATIVES BY THEIR PROVIDERS FOR PHARMACEUTICALS CLINICAL R&D DATA AVAILABILITY FACILITATES ANALYTICS AND INFORMATICS OPPORTUNITIES TO DEVELOP CURES AND ARREST OUTBREAKS FOR GOVERNMENT E-HEALTH CAN DRIVE HEALTH POLICIES, EARLY WARNING MECHANISMS AND ADVISORY MECHANISMS

Expert response #2	Code
The key stakeholders in my opinion are the government, healthcare providers, patients, pharmacies, laboratories and insurance companies.	GOVERNMENT HEALTHCARE PROVIDERS PATIENTS LABORATORIES INSURANCE COMPANIES

5. How can we encourage or ensure participation from these players?

Expert response #1	Code
Private insurance can ignite e-health by making it mandatory and paying for it through a compensatory mechanism (just like what Fed. Govt. does). In the long run, private insurances gain by efficiencies, prevention of fraud, waste and abuse by physicians. Patients insist on electronic health records that is single source and accessible by all his/her caregivers.	PRIVATE INSURERS CAN IGNITE E-HEALTH BY MAKING IT MANDATORY AND PAYING FOR IT THROUGH A COMPENSATORY MECHANISM LIKE THE FEDERAL GOVERNMENT DOES IN THE LONG RUN, PRIVATE INSURERS GAIN BY EFFICIENCIES, PREVENTION OF FRAUD, WASTE AND ABUSE BY PHYSICIANS PATIENTS SHOULD INSIST ON ELECTRONIC HEALTH RECORDS WHICH IS LONGITUDINAL AND ACCESSIBLE BY ALL PROVIDERS

Expert response #2	Code
The US government encourages e-health adoption by healthcare providers through incentives for adoption and penalties for non-adoption. The government's incentives are primarily aimed at covering the head-start costs and roll-over costs for an EHR system, and will come to an end after a healthcare provider achieves the 'meaningful use stage 3' criteria. Subsequent costs like the ongoing maintenance costs have to be borne by the healthcare providers. However by the time a healthcare provider meets the meaningful use stage 3 criteria, it would have achieved efficiencies which would translate to increased earnings, and which in turn, should help cover the maintenance costs. For instance, one such efficiency achieved is in terms of the average length of an appointment which has been found to have reduced from 45 to 15 minutes, enabling a doctor to see more patients /day, and thus generate more revenues.	THE GOVT. OFFERS INCENTIVES FOR ADOPTION AND PENALTIES FOR NON-ADOPTION. GOVT.'S INCENTIVES AIMED AT COVERING HEAD START COSTS AND ROLL OVER COSTS FOR AN EHR SYSTEM WILL COME TO AN END WHEN MEANINGFUL USE STAGE 3 CRITERION IS MET SUBSEQUENT COSTS LIKE ONGOING MAINTENANCE COSTS TO BE BORNE BY THE PROVIDERS BY THE TIME MEANINGFUL USE STAGE 3 CRITERIA ARE MET EFFICIENCIES WOULD BE ACHIEVED WHICH WOULD TRANSLATE TO INCREASED EARNINGS WHICH IN TURN SHOULD COVER MAINTENANCE COSTS

Appendix D

Individual healthcare providers (clinics) that are affiliated to Sutter Health will continue to incur subscription and support costs after the incentives are phased out. While some of these clinics attempt to recover their EHR-related costs by charging their patients for access to the EHR system, some others simply absorb it and give their patients free access to the system. Investing in and transitioning to e-health not only results in increased efficiencies, but also expands the network/market for the healthcare provider. However, a small number of individual healthcare providers find taking that step daunting because they are uncertain about the ROI. They prefer to incur the penalty associated with non-adoption which according to them may be far lower as compared to the long-term maintenance costs associated with an EHR system. Considering this deterring factor, it is not unfair for a clinic to charge its patients a reasonable amount for access to their heavily-invested in EHR system. In return, the patients can have their own EHR accounts, take ownership of their health information and control who they share their health information with.

The government on its part, can mandate filing back of health information into a cloud-based government system, and act as the custodian of its citizens' health information. This can be made possible if EHRs were open-source like the e-commerce systems.

FOR INSTANCE, LENGTH OF AN APPOINTMENT HAS BEEN FOUND TO HAVE REDUCED FROM 45 MINS. TO 15 MINS ENABLING A DOCTOR TO SEE MORE PATIENTS/DAY AND THUS GENERATE MORE REVENUES

INDIVIDUAL HEALTHCARE PROVIDERS (CLINICS) WILL CONTINUE TO INCUR SUBSCRIPTION AND SUPPORT COSTS AFTER THE INCENTIVES ARE PHASED OUT.

SOME OF THESE INDIVIDUAL PROVIDERS ATTEMPT TO RECOVER THEIR EHR-RELATED COSTS BY CHARGING THEIR PATIENTS FOR ACCESS TO THE SYSTEM AND SOME OTHERS JUST ABSORB THE COSTS

INVESTING IN E-HEALTH NOT ONLY LEADS TO INCREASED EFFICIENCIES BUT ALSO EXPANDS THE NETWORK/MARKET FOR THE PROVIDER

A SMALL NUMBER OF HEALTHCARE PROVIDERS FIND THE INVESTMENT DAUNTING DUE TO UNCERTAIN ROI. THEY PREFER TO INCUR PENALTIES FOR NON-ADOPTION WHICH IN THEIR OPINION MAY BE LOWER AS COMPARED TO THE LONG-TERM MAINTENANCE COSTS ASSOCIATED WITH EHR SYSTEMS

NOT UNFAIR FOR A CLINIC TO CHARGE ITS PATIENTS A REASONABLE AMOUNT FOR ACCESS TO THEIR HEAVILY INVESTED IN EHR SYSTEM

IN RETURN PATIENTS CAN HAVE THEIR OWN EHR ACCOUNTS, TAKE OWNERSHIP OF THEIR HEALTH INFORMATION AND CONTROL WHO THEY SHARE THEIR INFORMATION WITH

GOVT. CAN MANDATE FILING

| | BACK OF HEALTH INFORMATION INTO A CLOUD-BASED GOVT. SYSTEM AND ACT AS THE CUSTODIAN OF ITS CITIZEN'S HEALTH INFORMATION |
| | THIS IS POSSIBLE IF EHRS WERE OPEN-SOURCE LIKE THE E-COMMERCE SYSTEMS |

6. How will e-health benefit its various stakeholders and most importantly, the US?

Expert response #1	Code
At a high level, US scores poorly compared to peers on most health parameters despite multi fold budget. That correction is possible to some extent with e-health measures. The cost of a healthcare transaction is very high compared to similar transactions in other industries. Those wasted dollars can be channeled to research and coverage of indigent population. Obvious benefits to patients include better care, lesser number of repeat procedures and quicker determination of diagnosis.	US SCORES POORLY COMPARED TO PEERS ON MOST HEALTH PARAMETERS DESPITE MULTIFOLD BUDGET CORRECTION POSSIBLE WITH E-HEALTH MEASURES COST OF A HEALTHCARE TRANSACTION VERY HIGH COMPARED TO SIMILAR TRANSACTIONS IN OTHER INDUSTRIES. WASTED DOLLARS CAN BE CHANNELLED TO RESEARCH AND COVERAGE OF INDIGENT POPULATION
Whilst e-health may benefit most stakeholders, physicians may see pressure on their business. The clinical side will be efficient but that efficiency will cut into the high margins of the past in the business. For insurances and regulatory authority, this will mean lesser waste, fraud and abuse.	BENEFITS TO PATIENTS INCLUDE BETTER CARE, LESS NUMBER OF REPEAT PROCEDURES, QUICKER DETERMINATION OF DIAGNOSIS WHILE BENEFITS ACCRUE TO OTHER STAKEHOLDERS, PHYSICIANS FEEL PRESSURES. CLINICAL SIDE WILL BE EFFICIENT BUT EFFICIENCY WILL CUT INTO THE HIGH MARGINS OF THE PAST IN THE BUSINESS FOR INSURANCE COMPANIES AND REGULATORS, BENEFITS ARE LESS WASTE, FRAUD AND ABUSE

Expert response #2	Code
Healthcare providers gain efficiencies which translate into savings/increased revenues for them. E-visits and	HEALTHCARE PROVIDERS GAIN EFFICIENCIES THAT TRANSLATE INTO SAVINGS/INCREASED REVENUES FOR THEM

Appendix D 197

wireless communication between healthcare providers and patients are all made possible now. These advances have resulted in considerable cost and resource savings for healthcare providers. As for patients, they gain ubiquitous access to their health information via the EHR system that maintains their longitudinal health record. Developments like e-visits not only benefit patients and providers, but insurance companies also pay less for e-visits as compared to face-to-face consultations. The government has its population's health information readily accessible online and overall, e-health will lead to improvements in the population's health.	E-VISTS AND WIRELESS COMMUNICATION BETWEEN PROVIDERS AND PATIENTS POSSIBLE NOW ADVANCES HAVE RESULTED IN CONSIDERABLE COST AND RESOURCE SAVINGS FOR HEALTHCARE PROVIDERS PATIENTS GAIN UBIQUITOUS ACCESS TO THEIR HEALTH INFORMATION VIA EHR THAT MAINTAINS THEIR LONGITUDINAL HEALTH RECORD DEVELOPMENTS LIKE E-VISITS ALSO BENEFIT INSURERS AS THEY PAY LESS FOR E-VISITS AS COMPARED TO FACE-TO-FACE CONSULTATIONS GOVT. HAS ITS POPULATION'S HEALTH INFORMATION READILY ACCESSIBLE ONLINE OVERALL, E-HEALTH WILL LEAD TO IMPROVEMENTS IN POPULATION'S HEALTH

7. What is your opinion on the potential of cloud technology to transform healthcare in the US?

Expert response #1	Code
Cloud is an integral component of the e-health program. Several facets of e-health continue to depend on evolution of the cloud technology to be effective. High and easy availability, data porting and limitless storage enable e-health as is the lower cost of cloud storage. However some concerns regarding data safety and security continue to be held out against cloud storing sensitive health data.	CLOUD AN INTEGRAL COMPONENT OF THE E-HEALTH PROGRAM SEVERAL FACETS OF E-HEALTH CONTINUE TO DEPEND ON THE EVOLUTION OF THE CLOUD TECHNOLOGY TO BE EFFECTIVE HIGH AND EASY AVAILABILITY, DATA PORTING, LIMITLESS STORAGE, LOWER STORAGE COST ENABLE E-HEALTH CONCERNS REGARDING DATA SAFETY AND SECURITY CONTINUE TO BE HELD OUT AGAINST CLOUD STORING SENSITIVE DATA

Expert response #2	Code
There are no major players on cloud yet. Government policies and grants to support this direction may encourage healthcare providers to embrace cloud technology in the future. Ten years down the line, it is likely that all health information may be on cloud considering the exponential rate at which health information is growing. Security and confidentiality of data are of course key concerns with respect to the cloud technology, but these can be overcome with the help of encryption technologies which are ever-evolving. The government should assume the responsibility to guard the e-health system to maintain public confidence - similar to what they have been doing with the national financial system through the FDIC (Federal Deposit Insurance System).	NO MAJOR PLAYERS ON CLOUD YET GOVT. POLICIES AND GRANTS TO SUPPORT THIS DIRECTION MAY ENCOURAGE HEALTHCARE PROVIDERS TO EMBRACE THE CLOUD TECHNOLOGY IN FUTURE 10 YEARS DOWN THE LINE IT IS LIKELY THAT ALL HEALTH INFORMATION MAY BE ON CLOUD CONSIDERING THE EXPONENTIAL RATE AT WHICH HEALTH INFORMATION IS GROWING SECURITY AND CONFIDENTIALITY OF DATA ARE KEY CONCERNS W.R.T. CLOUD. CAN BE OVERCOME WITH ENCRYPTON TECHNOLOGIES WHICH ARE EVER-EVOLVING GOVT. SHOULD ASSUME THE RESPONSIBILITY TO GUARD THE E-HEALTH ECOSYSTEM TO MAINTAIN PUBLIC CONFIDENCE – SIMILAR TO WHAT THEY HAVE BEEN DOING WITH THE NATIONAL FINANCIAL SYSTEM THROUGH THE FDIC (FEDERAL DEPOSIT INSURANCE SYSTEM)

References

2015 Health Care Outlook India. (2015). Retrieved 8 November 2016, from https://www2. deloitte.com/content/dam/Deloitte/global/Documents/Life-Sciences-Health-Care/gx-lshc-2015-health-care-outlook-india.pdf

About the Precision Public Health Summit (2016). University of California, San Francisco. https://precisionmedicine.ucsf.edu/about-precision-public-health-summit

Abukhousa, E., Mohamed, N. and Al-Jaroodi, J. (2012). e-Health cloud: Opportunities and challenges. *Future Internet*, 4(3), 621-645. doi:10.3390/fi4030621

Acampora, G., Cook, D.J., Rashidi, P. and Vasilakos, A.V. (2013). A survey on ambient intelligence in health care. *Proceedings of the IEEE. Institute of Electrical and Electronics Engineers*, 101(12), 2470-2494. https://doi.org/10.1109/JPROC.2013.2262913

Accenture Healthcare Consumer Survey. (2012). Do Healthcare Consumers Want One-Stop Shopping for Care and Coverage? Retrieved 8 March 2015, from https://www. accenture.com/sg-en/~/media/Accenture/Conversion-Assets/DotCom/Documents/ Global/PDF/Industries_11/Accenture-Do-Healthcare-Consumers-Want-One-Stop-Shopping-for-Care-and-Coverage.pdf

Accenture, (2012). Connected health: The drive to integrated healthcare delivery. *Accenture*. Retrieved from http://nstore.accenture.com/acn_com/PDF/Accenture-Connected-Health-Global-Report-Final-Web.pdf

Adar, E. and Huberman, B. (2000). Free riding on Gnutella. *First Monday*, 5(10), http://dx.doi.org/10.5210/fm.v5i10.792

Adenuga, K.I., Adenuga, R.O., Ziraba, A. and Mbuh, P.E. (2019). Healthcare augmentation: Social adoption of augmented reality glasses in medicine. *Proceedings of the 2019 8th International Conference on Software and Information Engineering - ICSIE '19*, 71-74. https://doi.org/10.1145/3328833.3328840

Adler-Milstein, J. and Bates, D. (2010). Paperless healthcare: Progress and challenges of an IT-enabled healthcare system. *Business Horizons*, 53(2), 119-130.

Adler-Milstein, J., Bates, D. and Jha, A. (2011). A survey of health information exchange organizations in the United States: Implications for meaningful use. *Annals of Internal Medicine*, 154(10), 666. http://dx.doi.org/10.7326/0003-4819-154-10-201105170-00006

Agbo, C., Mahmoud, Q. and Eklund, J. (2019). Blockchain technology in healthcare: A systematic review. *Healthcare*, 7(2), 56. https://doi.org/10.3390/healthcare7020056

Aggrawal, A.K. and Travers, S. (2001). e-Commerce in healthcare: Changing the traditional landscape. *Journal of Healthcare Information Management*, 15(1), 25-36.

Ahern, D.K. (2007). Challenges and opportunities of e-Health research. *American Journal of Preventive Medicine*, 32(5S), S75-S82.

Ahern, D., Woods, S., Lightowler, M., Finley, S. and Houston, T. (2011). Promise of and potential for patient-facing technologies to enable meaningful use. *American Journal of Preventive Medicine*, 40(5), S162-S172. http://dx.doi.org/10.1016/j.amepre.2011.01.005

Ahram, T., Sargolzaei, A., Sargolzaei, S., Daniels, J. and Amaba, B. (2017). Blockchain technology innovations. *2017 IEEE Technology Engineering Management Conference (TEMSCON)*, 137-141. https://doi.org/10.1109/TEMSCON.2017.7998367

Al Omar, A., Rahman, M.S., Basu, A. and Kiyomoto, S. (2017). MediBchain: A blockchain based privacy preserving platform for healthcare data. pp. 534-543. *In*: G. Wang, M. Atiquzzaman, Z. Yan and K.-K.R. Choo (Eds.), Security, Privacy, and Anonymity in Computation, Communication, and Storage. Springer International Publishing. https://doi.org/10.1007/978-3-319-72395-2_49

Alaa, A.M., Moon, K.H., Hsu, W. and Van Der Schaar, M. (2016). Confident care: A clinical decision support system for personalized breast cancer screening. *IEEE Transactions on Multimedia*, 18(10), 1942-1955. Scopus. https://doi.org/10.1109/TMM.2016.2589160

Alaraj, A., Luciano, C.J., Bailey, D.P., Elsenousi, A., Roitberg, B.Z., Bernardo, A., Banerjee, P.P. and Charbel, F.T. (2015). Virtual reality cerebral aneurysm clipping simulation with real-time haptic feedback. *Neurosurgery*, 11(Suppl 2), 52-58. https://doi.org/10.1227/NEU.0000000000000583

Albanese, R. and Van Fleet, D. (1985). Rational behavior in groups: The free-riding tendency. *Academy of Management Review*, 10(2), 244-255. http://dx.doi.org/10.5465/amr.1985.4278118

Alfian, G., Syafrudin, M., Ijaz, M.F., Syaekhoni, M.A., Fitriyani, N.L. and Rhee, J. (2018). A personalized healthcare monitoring system for diabetic patients by utilizing BLE-based sensors and real-time data processing. *Sensors* (Switzerland), 18(7). Scopus. https://doi.org/10.3390/s18072183

Alketbi, A., Nasir, Q. and Talib, M.A. (2018). Blockchain for government services—Use cases, security benefits and challenges. *2018 15th Learning and Technology Conference (L T)*, 112-119. https://doi.org/10.1109/LT.2018.8368494

Almogren, A. (2018). An automated and intelligent Parkinson disease monitoring system using wearable computing and cloud technology. *Cluster Computing*, 1-8. Scopus.

Alonso, S.G., de la Torre Díez, I., Rodrigues, J.J.P.C., Hamrioui, S. and López-Coronado, M. (2017). A systematic review of techniques and sources of Big Data in the healthcare sector. *Medical Systems*, 41(11), 183. https://doi.org/10.1007/s10916-017-0832-2

Aloqaily, M., Otoum, S., Ridhawi, I.A. and Jararweh, Y. (2019). An intrusion detection system for connected vehicles in smart cities. *Ad Hoc Networks*, 90, 101842. https://doi.org/10.1016/j.adhoc.2019.02.001

Altini, M., Penders, J., Vullers, R. and Amft, O. (2015). Estimating energy expenditure using body-worn accelerometers: A comparison of methods, sensors number and positioning. *IEEE Journal of Biomedical and Health Informatics*, 19(1), 219-226. https://doi.org/10.1109/JBHI.2014.2313039

Amatayakul, M.K. (2009). *Electronic Health Records: A Practical Guide for Professionals and Organizations*. Chicago, IL: American Health Information Management Association.

Amin, P., Anikireddypally, N.R., Khurana, S., Vadakkemadathil, S. and Wu, W. (2019). *Personalized Health Monitoring Using Predictive Analytics*. 271-278. Scopus. https://doi.org/10.1109/BigDataService.2019.00048

References

Andoni, M., Robu, V., Flynn, D., Abram, S., Geach, D., Jenkins, D., McCallum, P. and Peacock, A. (2019). Blockchain technology in the energy sector: A systematic review of challenges and opportunities. *Renewable and Sustainable Energy Reviews*, 100, 143-174. https://doi.org/10.1016/j.rser.2018.10.014

Androulaki, E., Barger, A., Bortnikov, V., Cachin, C., Christidis, K., De Caro, A., Enyeart, D., Ferris, C., Laventman, G., Manevich, Y., Muralidharan, S., Murthy, C., Nguyen, B., Sethi, M., Singh, G., Smith, K., Sorniotti, A., Stathakopoulou, C., Vukolić, M., ... Yellick, J. (2018). Hyperledger fabric: A distributed operating system for permissioned blockchains. *Thirteenth EuroSys Conference*, 1-15. https://doi.org/10.1145/3190508.3190538

Anoshiravani, A., Gaskin, G., Kopetsky, C.S. and Longhurst, C.A. (2011). Implementing an interoperable personal health record in pediatrics: Lessons learned at an academic children's hospital. *Journal of Participatory Medicine*, 3.

Antonopoulos, A.M. (2017). *Mastering Bitcoin: Programming the Open Blockchain* (2nd ed.). O'Reilly Media.

Archer, N., Fevrier-Thomas, U., Lokker, C., McKibbon, K.A. and Straus, S.E. (2011). Personal health records: A scoping review. *J. Am. Med. Inform. Assoc.*, 18(4), 515-522. doi:10.1136/amiajnl-2011-000105

Armbrust, M., Fox, A., Griffith, R., Joseph, A.D., Katz, R.H., Konwinski, A., Lee, G., Patterson, D.A., Rabkin, A., Stoica, I. and Zaharia, M. (2009). Above the Clouds: A Berkeley View of Cloud Computing. *Electrical Engineering and Computer Sciences*. University of California at Berkeley. Retrieved from http://www.eecs.berkeley.edu/Pubs/TechRpts/2009/EECS-2009-28.pdf

Arrow, K.J. (1963). Uncertainty and the welfare economics of medical care. *The American Economic Review*, LIII(5), 141-149.

Ashbrook, D. and Starner, T. (2003). Using GPS to learn significant locations and predict movement across multiple users. *Personal and Ubiquitous Computing*, 7(5), 275-286. https://doi.org/10.1007/s00779-003-0240-0

Atkinson, N.L., Saperstein, S.L. and Pleis, J. (2009). Using the Internet for health-related activities: Findings from a national probability sample. *Journal of Medical Internet Research*, 11(1), e4.

Awad, M. and Khanna, R. (2015). Machine learning. pp. 1-18. *In*: M. Awad and R. Khanna (Eds.), Efficient Learning Machines: Theories, Concepts, and Applications for Engineers and System Designers. Apress. https://doi.org/10.1007/978-1-4302-5990-9_1

Azaria, A., Ekblaw, A., Vieira, T. and Lippman, A. (2016). MedRec: Using blockchain for medical data access and permission management. *2016 2nd International Conference on Open and Big Data (OBD)*, 25-30. https://doi.org/10.1109/OBD.2016.11

Baek, J.-W., Jung, H. and Chung, K. (2019). Context mining based mental health model for lifecare platform. *Medico-Legal Update*, 19(1), 674-679. Scopus. https://doi.org/10.5958/0974-1283.2019.00119.1

Bahga, A. and Madisetti, V.K. (2013). A cloud-based approach for interoperable electronic health records (EHRs). *IEEE Journal of Biomedical and Health Informatics*, 17(5), 894-906.

Bahrain News Agency | Kingdom of Bahrain wins two major awards in Seoul. (2014). Bna. bh., from http://bna.bh/portal/en/news/623813?date=2014-07-7

Bali, M., Mohanty, S., Chatterjee, S., Sarma, M. and Puravankara, R. (2019). Diabot: A predictive medical chatbot using ensemble learning. *Recent Technology and Engineering*, 8(2), 6334-6340. Scopus. https://doi.org/10.35940/ijrte.B2196.078219

Bandyopadhyay. S., Ozdemir, Z. and Barron, J. (2012). The future of personal health records in the presence of misaligned incentives. *Communications of the Association for Information Systems*, 31, 155-166.

Banerjee, M., Lee, J. and Choo, K.-K.R. (2018). A blockchain future for internet of things security: A position paper. *Digital Communications and Networks*, 4(3), 149-160. https://doi.org/10.1016/j.dcan.2017.10.006

Banos, O., Garcia, R., Holgado-Terriza, J.A., Damas, M., Pomares, H., Rojas, I., Saez, A. and Villalonga, C. (2014). mHealthDroid: A novel framework for agile development of mobile health applications. pp. 91-98. *In*: L. Pecchia, L.L. Chen, C. Nugent and J. Bravo (Eds.), Ambient Assisted Living and Daily Activities. Springer.

Banos, O., Moral-Munoz, J.A., Diaz-Reyes, I., Arroyo-Morales, M., Damas, M., Herrera-Viedma, E., Hong, C.S., Lee, S., Pomares, H., Rojas, I. and Villalonga, C. (2015). mDurance: A novel mobile health system to support trunk endurance assessment. *Sensors* (Basel, Switzerland), 15(6), 13159-13183. https://doi.org/10.3390/s150613159

Baro, E., Degoul, S., Beuscart, R. and Chazard, E. (2015). Toward a literature-driven definition of Big Data in healthcare [Research article]. *BioMed Research International*. https://doi.org/10.1155/2015/639021

Bassi, J. and Lau, F. (2013). Measuring value for money: A scoping review on economic evaluation of health information systems. *Journal of the American Medical Informatics Association*, 20, 792-801.

Batchelor, J. (2009). Future Forecast: Cloud Computing Brightens Healthcare's Dark Skies. *CMIO*. Retrieved 6 May 2009, from http://www.clinical-innovation.com/topics/technology-management/future-forecast-cloud-computing-brightens-healthcares-dark-skies

Baur, A., Fehr, J., Mayer, C., Pawlu, C. and Schaudel, F. (2011). Health care beyond medicine: Meeting the demand for new forms of care. *Health International*, 56-63. Retrieved from http://www.mckinsey.com/search.aspx?q=Health+care+beyond+medicine

Baur, C. (2008). An analysis of factors underlying e-Health disparities. *Cambridge Quarterly of Healthcare Ethics*, 17, 417-428.

Beck, R., Avital, M., Rossi, M. and Thatcher, J.B. (2017). Blockchain technology in business and information systems research. *Business & Information Systems Engineering*, 59(6), 381-384. https://doi.org/10.1007/s12599-017-0505-1

Benkler, Y. (2006). *The Wealth of Networks* (1st ed.). New Haven [Conn.]: Yale University Press.

Bergmo, T.S. (2015). How to measure costs and benefits of e-Health interventions: An overview of methods and frameworks. *J. Med. Internet. Res.*, 17(11), e254. doi:10.2196/jmir.4521

Berthold, J. (2008). Investing in EHRs pays off in paperless perks. *ACP Internist*. Retrieved from http://www.acpinternist.org/archives/2008/02/ehrs.htm

Bianchi, F., Redmond, S.J., Narayanan, M.R., Cerutti, S. and Lovell, N.H. (2010). Barometric pressure and triaxial accelerometry-based falls event detection. *IEEE Transactions on Neural Systems and Rehabilitation Engineering*, 18(6), 619-627. https://doi.org/10.1109/TNSRE.2010.2070807

Billis, A.S., Papageorgiou, E.I., Frantzidis, C.A., Tsatali, M.S., Tsolaki, A.C. and Bamidis, P.D. (2015). A decision-support framework for promoting independent living and ageing well. *IEEE Journal of Biomedical and Health Informatics*, 19(1), 199-209. https://doi.org/10.1109/JBHI.2014.2336757

References

Black, A., Car, J., Pagliari, C., Anandan, C., Cresswell, K. and Bokun, T. et al. (2011). The impact of e-Health on the quality and safety of health care: A systematic overview. *Plos Med*, 8(1), e1000387. http://dx.doi.org/10.1371/journal.pmed.1000387

Blackwell, G. (2008). The future of IT in healthcare. *Informatics for Health and Social Care*, 33(4), 211-326.

Blagojevic, D. (2019, March 21). What is practical Byzantine Fault Tolerance (pBFT)? *CaptainAltcoin.* https://captainaltcoin.com/what-is-practical-byzantine-fault-tolerance-pbft/

Blank, R. (2012). Transformation of the US Healthcare System: Why is change so difficult? *Current Sociology*, 60(4), 415-426. http://dx.doi.org/10.1177/0011392112438327

Blomqvist, A. (1991). The doctor as double agent: Information asymmetry, health insurance and medical care. *Journal of Health Economics*, 10(1991), 411-432.

Blumenthal, D. (2009). Stimulating the adoption of health information technology. *New England Journal of Medicine*, 360(15), 1477-1479.

Blumenthal, D. and Tavenner, M. (2010). The "Meaningful Use" regulation for electronic health records. *New England Journal of Medicine*, 363(6), 501-504.

Boneh, D., Di Crescenzo, G., Ostrovsky, R. and Persiano, G. (2004). Public key encryption with keyword search. pp. 506-522. *In*: C. Cachin and J.L. Camenisch (Eds.), Advances in Cryptology—EUROCRYPT 2004. Springer. https://doi.org/10.1007/978-3-540-24676-3_30

Bowen, G.A. (2006). Grounded theory and sensitizing concepts. *International Journal of Qualitative Methods*, 5(3), 12-23.

Boyd, H.W., Westfall, R. and Stasch, S.F. (1994). *Marketing Research: Text and Cases* (7th ed.). Homewood, IL: Richard D. Irwin, Inc.

Brailer, D.J. (2005). Interoperability: The key to the future health care system. *Health Affairs-Millwood VA then Bethesda MA-*, 24, W5.

Brandenburger, A. and Nalebuff, B. (1996). *Co-opetition.* New York: Doubleday.

Brandenburger, A.M. and Nalebuff, B.J. (1995). The right game: Use game theory to shape strategy. *Harvard Business Review*, 73(4), 57-71.

Brandenburger, A.M. and Stuart, H.W. (1996). Value-based business strategy. *Journal of Economics and Management Strategy*, 5(1), 5-24.

Brandt, P., Basten, T., Stuiik, S., Bui, V., de Clercq, P., Pires, L.F. and van Sinderen, M. (2013). Semantic interoperability in sensor applications making sense of sensor data. *2013 IEEE Symposium on Computational Intelligence in Healthcare and E-Health (CICARE)*, 34-41. https://doi.org/10.1109/CICARE.2013.6583065

Bresnick, J. (2017a, July 28). *IBM Watson, Quest Launch Genomic Cognitive Computing Partnership.* HealthITAnalytics. https://healthitanalytics.com/news/ibm-watson-quest-launch-genomic-cognitive-computing-partnership

Bresnick, J. (2017b, October 19). *NIH, Pharma Orgs Launch $215M Precision Medicine, Cancer Project.* HealthITAnalytics. https://healthitanalytics.com/news/nih-pharma-orgs-launch-215m-precision-medicine-cancer-project

Bresnick, J. (2018a, January 9). *GE, Roche Partner for Big Data Analytics, Precision Medicine Platform.* HealthITAnalytics. https://healthitanalytics.com/news/ge-roche-partner-for-big-data-analytics-precision-medicine-platform

Bresnick, J. (2018b, January 11). *What Are Precision Medicine and Personalized Medicine?* HealthITAnalytics. https://healthitanalytics.com/features/what-are-precision-medicine-and-personalized-medicine

Britnell, M. (2015). Transforming health care takes continuity and consistency. *Harvard*

Business Review. https://hbr.org/2015/12/transforming-health-care-takes-continuity-and-consistency

Broderick, M. and Smaltz, D.H. (2003). HIMSS e-health white paper: e-health defined. Chicago: Healthcare Information and Management Systems Society.

Brynjolfsson, E. and Hitt, L.M. (1998). Beyond the productivity paradox. *Communications of the ACM*, 41(8), 49-55.

Brytskyi, O. (2018, September 26). *Decentralized AI: Blockchain's bright future*. https://espeoblockchain.com/blog/decentralized-ai-benefits/

Bulgiba, A.M. (2004). Information Technology in Health Care – What the Future Holds. *Asia-Pacific Journal of Public Health*, 16(1), 64-71.

Buntin, M., Burke, M., Hoaglin, M. and Blumenthal, D. (2011). The benefits of health information technology: A review of the recent literature shows predominantly positive results. *Health Affairs*, 30(3), 464-471. http://dx.doi.org/10.1377/hlthaff.2011.0178

Burkhard, R. (2009). Book review: Patient-centred e-health (Review of the book Patient-Centred E-Health). *International Journal of E-Services and Mobile Applications*, 1(3), 79-81.

Burnard, P. (1991). A method of analysing interview transcripts in qualitative research. *Nurse Education Today*, 11(6), 461-466.

Busch, R.S. (2008). *Electronic Health Records: An Audit and Internal Control Guide*. New Jersey: John Wiley & Sons, Inc.

Business Wire. (2016). Verisk Health Announces Rebrand to Verscend Technologies. Businesswire.com. Retrieved 29 September 2016, from http://www.businesswire.com/news/home/20160826005091/en/Verisk-Health-Announces-Rebrand-Verscend-Technologies

Cabitza, F., Simone, C. and Cornetta, D. (2015). Sensitizing concepts for the next community-oriented technologies: Shifting focus from social networking to convivial artifacts. *The Journal of Community Informatics*, 11(2).

Cai, C.W. (2018). Disruption of financial intermediation by FinTech: A review on crowdfunding and blockchain. *Accounting & Finance*, 58(4), 965-992. https://doi.org/10.1111/acfi.12405

Calvillo, J., Román, I. and Roa, L.M. (2013). How technology is empowering patients: A literature review. *Health Expectations,* doi: 10.1111/hex.12089

Carey, D.J., Fetterolf, S.N., Davis, F.D., Faucett, W.A., Kirchner, H.L., Mirshahi, U., Murray, M.F., Smelser, D.T., Gerhard, G.S. and Ledbetter, D.H. (2016). The Geisinger MyCode community health initiative: An electronic health record-linked biobank for precision medicine research. *Genetics in Medicine*, 18(9), 906-913. https://doi.org/10.1038/gim.2015.187

Castro, D., Coral, W., Rodriguez, C., Cabra, J. and Colorado, J. (2017). Wearable-based human activity recognition using an IoT Approach. *Journal of Sensor and Actuator Networks*, 6(4). Scopus. https://doi.org/10.3390/jsan6040028

Castro, M. and Liskov, B. (2002). Practical byzantine fault tolerance and proactive recovery. *ACM Transactions on Computer Systems*, 20(4), 398-461. https://doi.org/10.1145/571637.571640

Catalino, J. (2010). Software solutions can trim rising costs. *Health Management Technology*, 31(3), 10-11.

Centre for Medicare and Medicaid Services (2016). Retrieved from https://www.cms.gov/

Chang, H.H., Chou, P.B. and Ramakrishnan, S. (2009). An Ecosystem Approach for Healthcare Services Cloud. pp. 608-612. IEEE International Conference on e-Business Engineering (ICEBE '09).

References

Chang, R. (2010). 2 health-care clusters not enough. *The Straits Times*. Retrieved from https://www.healthxchange.com.sg/News/Pages/Two-healthcare-clusters-not-enough.aspx

Charette, R. (2006). EHRs: Electronic Health Records or Exceptional Hidden Risks. *Communications of the ACM*, 49(6), 120.

Charles, D., Gabriel, M. and Searcy, T. (April 2015). Adoption of Electronic Health Record Systems among U.S. Non-Federal Acute Care Hospitals: 2008-2014. ONC Data Brief, no. 23. Office of the National Coordinator for Health Information Technology: Washington DC.

Charmaz, K. (2003). Grounded theory: Objectivist and constructivist methods. pp. 249-291. *In*: N.K. Denzin and Y.S. Lincoln (Eds.), Strategies for Qualitative Inquiry (2nd ed.). Thousand Oaks, CA: Sage.

Charmaz, K. (2014). *Constructing Grounded Theory (*2nd ed.*)*. London: Sage Publications.

Chaudhry, B., Wang, J., Wu, S., Maglione, M., Mojica, W., Roth, E., . . . Shekelle, P.G. (2006). Systematic review: Impact of health information technology on quality, efficiency, and costs of medical care. *Ann Intern Med*, 144(10), 742-752.

Chen, J. and Zheng, X. (2015). A system architecture for smart health services and applications. *Lecture Notes in Computer Science* (Including Subseries Lecture Notes in Artificial Intelligence and Lecture Notes in Bioinformatics), 9426, 449-456. Scopus. https://doi.org/10.1007/978-3-319-26181-2_42

Chen, K.-H., Chen, P.-C., Liu, K.-C. and Chan, C.-T. (2015). Wearable sensor-based rehabilitation exercise assessment for knee osteoarthritis. *Sensors* (Basel, Switzerland), 15(2), 4193-4211. https://doi.org/10.3390/s150204193

Chen, S. and Wong, S. (2014). Singapore Beats Hong Kong in Health Efficiency: Southeast Asia. Bloomberg.com. Retrieved from http://www.bloomberg.com/news/articles/2014-09-18/singapore-beats-hong-kong-in-health-efficiency-southeast-asia

Chen, Y., Ding, S., Xu, Z., Zheng, H. and Yang, S. (2018). Blockchain-based medical records secure storage and medical service framework. *Medical Systems*, 43(1), 5. https://doi.org/10.1007/s10916-018-1121-4

Chesbrough, H. and Rosenbloom, R.S. (2002). The role of the business model in capturing value from innovation: Evidence from Xerox Corporation's technology. *Industrial and Corporate Change*, 11(3), 529-555.

Chia, A.-R., de Seymour, J.V., Colega, M., Chen, L.-W., Chan, Y.-H., Aris, I.M., Tint, M.-T., Quah, P.L., Godfrey, K.M., Yap, F., Saw, S.-M., Baker, P.N., Chong, Y.-S., van Dam, R.M., Lee, Y.S. and Chong, M.F.-F. (2016). A vegetable, fruit, and white rice dietary pattern during pregnancy is associated with a lower risk of preterm birth and larger birth size in a multiethnic Asian cohort: The Growing Up in Singapore Towards healthy Outcomes (GUSTO) cohort study. *Clinical Nutrition*, 104(5), 1416-1423. https://doi.org/10.3945/ajcn.116.133892

Chin, R. (2000). The Internet: Another facet to the paradigm shift in healthcare. *Singapore Medical Journal*, 41(9), 426-430.

Chitchyan, R. and Murkin, J. (2018). Review of blockchain technology and its expectations: Case of the energy sector. *ArXiv:1803.03567 [Cs]*. http://arxiv.org/abs/1803.03567

Chiuchisan, I., Costin, H.-N. and Geman, O. (2014). Adopting the Internet of Things technologies in health care systems. *2014 International Conference and Exposition on Electrical and Power Engineering (EPE)*, 532-535. https://doi.org/10.1109/ICEPE.2014.6969965

Choi, S., Jung, K. and Noh, S.D. (2015). Virtual reality applications in manufacturing industries: Past research, present findings, and future directions. *Concurrent Engineering*, 23(1), 40-63. https://doi.org/10.1177/1063293X14568814

Choudhury, T., Borriello, G., Consolvo, S., Haehnel, D., Harrison, B., Hemingway, B., Hightower, J., Klasnja, P. "Pedja", Koscher, K., LaMarca, A., Landay, J.A., LeGrand, L., Lester, J., Rahimi, A., Rea, A. and Wyatt, D. (2008). The mobile sensing platform: An embedded activity recognition system. *IEEE Pervasive Computing*, 7(2), 32-41. https://doi.org/10.1109/MPRV.2008.39

Christensen, C. (1997). *The Innovator's Dilemma* (1st ed.). Boston, Mass.: Harvard Business School Press.

Chung, P.-C. and Liu, C.-D. (2008). A daily behavior enabled hidden Markov model for human behavior understanding. *Pattern Recognition*, 41(5), 1572-1580. https://doi.org/10.1016/j.patcog.2007.10.022

Cichosz, S.L., Stausholm, M.N., Kronborg, T., Vestergaard, P. and Hejlesen, O. (2019). How to use blockchain for diabetes health care data and access management: An operational concept. *Diabetes Science and Technology*, 13(2), 248-253. https://doi.org/10.1177/1932296818790281

Clancy , C.M., Anderson, K.M. and White, P.J. (2009). Investing in health information infrastructure: Can it help achieve health reform? *Health Affairs*, 28(2), 478-482.

Clarke III, I., Flaherty, T.B., Hollis, S.M. and Tomallo, M. (2009). Consumer privacy issues associated with the use of electronic health records. *Academy of Health Care Management Journal*, 5(1/2), 63.

Clarke, J.L. and Meiris, D.C. (2006). Electronic personal health records come of age. *American Journal of Medical Quality*, 21(3 suppl), 5S-15S. http://dx.doi.org/10.1177/1062860606287642

Codreanu, I.A. and Florea, A.M. (2016). *A Proposed Serious Game Architecture to Self-Management HealthCare for Older Adults*. 437-440. Scopus. https://doi.org/10.1109/SYNASC.2015.71

Comandé, G., Nocco, L. and Peigné, V. (2015). An empirical study of healthcare providers and patients' perceptions of electronic health records. *Computers in Biology and Medicine*, 59, 194-201. http://dx.doi.org/10.1016/j.compbiomed.2014.01.011

Connell, J. (2013). Contemporary medical tourism: Conceptualisation, culture and commodification. *Tourism Management*, 34, 1-13. http://dx.doi.org/10.1016/j.tourman.2012.05.009

Conoscenti, M., Vetro, A. and De Martin, J.C. (2016). Blockchain for the Internet of Things: A systematic literature review. *13th International Conference of Computer Systems and Applications (AICCSA)*, 1-6. https://doi.org/10.1109/AICCSA.2016.7945805

Conti, M., Sandeep Kumar, E., Lal, C. and Ruj, S. (2018). A survey on security and privacy issues of bitcoin. *IEEE Communications Surveys Tutorials*, 20(4), 3416-3452. https://doi.org/10.1109/COMST.2018.2842460

Corbin, J. and Strauss, A. (1990). Grounded theory research: Procedures, canons, and evaluative criteria. *Qualitative Sociology*, 13(1), 3-21. http://dx.doi.org/10.1007/bf00988593

Cordina, J., Kumar, R. and Moss, C. (2015). Debunking common myths about healthcare consumerism. McKinsey & Company. Retrieved from http://www.mckinsey.com/industries/healthcare-systems-and-services/our-insights/debunking-common-myths-about-healthcare-consumerism

Couch, J.B. (2013). How to use electronic health records to achieve the quadruple aim. *JAMA Internal Medicine*, 173(3), 250-250.

Creswell, J. (2003). *Research Design: Qualitative, Quantitative and Mixed Methods Approaches* (2nd ed.). Thousand Oaks, CA: Sage Publications.

Croman, K., Decker, C., Eyal, I., Gencer, A.E., Juels, A., Kosba, A., Miller, A., Saxena, P., Shi, E., Gün Sirer, E., Song, D. and Wattenhofer, R. (2016). On scaling decentralized blockchains. pp. 106-125. *In*: J. Clark, S. Meiklejohn, P.Y.A. Ryan, D. Wallach, M. Brenner and K. Rohloff (Eds.), Financial Cryptography and Data Security. Springer. https://doi.org/10.1007/978-3-662-53357-4_8

Cromwell, J., Trisolini, M.G., Pope, G.C., Mitchell, J.B. and Greenwald, L.M. (Eds). (2011). *Pay for Performance in Health Care: Methods and Approaches*. RTI Press Publication No. BK-0002-1103. Research Triangle Park, NC: RTI Press. Retrieved from http://www.rti.org/rtipress.

Cruz, M.F., Cavalcante, C.A.M.T. and Sá Barretto, S.T. (2018). Using OPC and HL7 Standards to incorporate an industrial Big Data historian in a health IT environment. *Medical Systems*, 42(7), 122. https://doi.org/10.1007/s10916-018-0979-5

Cummings, E. (2015). Understanding the U.S. Health Care System. InterExchange. Retrieved from https://www.interexchange.org/articles/abroad/2015/08/10/understanding-us-healthcare-system/

Dagher, G.G., Mohler, J., Milojkovic, M. and Marella, P.B. (2018). Ancile: Privacy-preserving framework for access control and interoperability of electronic health records using blockchain technology. *Sustainable Cities and Society*, 39, 283-297. https://doi.org/10.1016/j.scs.2018.02.014

Daowd, A., Abidi, S.R., Abusharekh, A. and Raza Abidi, S.S. (2018). A personalized risk stratification platform for population lifetime healthcare. *Studies in Health Technology and Informatics*, 247, 920-924. Scopus. https://doi.org/10.3233/978-1-61499-852-5-920

Dash, P. and Meredith, D. (2010). When and how provider competition can improve health care delivery. *McKinsey Quarterly*. https://www.mckinsey.com/industries/healthcare-systems-and-services/our-insights/when-and-how-provider-competition-can-improve-health-care-delivery#

Dash—Dash is Digital Cash You Can Spend Anywhere. (n.d.). Dash. Retrieved October 25, 2019, from https://www.dash.org/

Datta, S.K., Bonnet, C., Gyrard, A., Ferreira da Costa, R.P. and Boudaoud, K. (2015). Applying Internet of Things for personalized healthcare in smart homes. *24th Wireless and Optical Communication Conference (WOCC)*, 164-169. https://doi.org/10.1109/WOCC.2015.7346198

Davis, K., Stremikis, K., Schoen, C. and Squires, D. (2014). Mirror, mirror on the wall, 2014 update: How the US health care system compares internationally. *The Commonwealth Fund*, 16.

de Brantes, F., Emery, D.W., Overhage, J.M., Glaser, J. and Marchibroda, J. (2007). The potential of HIEs as Infomediaries. *Journal of Healthcare Information Management*, 21(1), 69-75.

de Grey, A.D.N.J. (2016). Artificial Intelligence and Medical Research: Time to Aim Higher? *Rejuvenation Research*, 19(2), 105-106. https://doi.org/10.1089/rej.2016.1827

De Silva, D., Ranasinghe, W., Bandaragoda, T., Adikari, A., Mills, N., Iddamalgoda, L., Alahakoon, D., Lawrentschuk, N., Persad, R., Osipov, E., Gray, R. and Bolton, D. (2018). Machine learning to support social media empowered patients in cancer care and cancer treatment decisions. *PLoS ONE*, 13(10). Scopus. https://doi.org/10.1371/journal.pone.0205855

Debarba, H.G., De Oliveira, M.E., Ladermann, A., Chague, S. and Charbonnier, C. (2018a). Augmented reality visualization of joint movements for physical examination and rehabilitation. *Virtual Reality and 3D User Interfaces*, 537-538. Scopus. https://doi.org/10.1109/VR.2018.8446368

Debarba, H.G., De Oliveira, M.E., Ladermann, A., Chague, S. and Charbonnier, C. (2018b). Augmented reality visualization of joint movements for rehabilitation and sports medicine. *20th Symposium on Virtual and Augmented Reality*, 114-121. Scopus. https://doi.org/10.1109/SVR.2018.00027

DeNardis, L. (2014). E-health standards and interoperability. Retrieved from https://itunews.itu.int/En/2472-E8209;health-standards-and-interoperability.note.aspx

DesRoches, C., Campbell, E., Vogeli, C., Zheng, J., Rao, S. and Shields, A. et al. (2010). Electronic Health Records' Limited Successes Suggest More Targeted Uses. *Health Affairs*, 29(4), 639-646. http://dx.doi.org/10.1377/hlthaff.2009.1086

Devaraj, S. and Kohli, R. (2000). Information technology payoff in the health-care industry: A longitudinal study. *Journal of Management Information Systems*, 16(4), 41-67.

DeVore, S. and Champion, R. (2011). Driving population health through accountable care organizations. *Health Affairs*, 30(1), 41-50. http://dx.doi.org/10.1377/hlthaff.2010.0935

Diamond, C., Mostashari, F. and Shirky, C. (2009). Collecting and sharing data for population health: A new paradigm. *Health Affairs*, 28(2), 454-466. http://dx.doi.org/10.1377/hlthaff.28.2.454

Digital.Transformation of Industries: Healthcare (p. 36). (2016). World Economic Forum. http://reports.weforum.org/digital-transformation/wp-content/blogs.dir/94/mp/files/pages/files/wef-dti-healthcarewhitepaper-final-january-2016.pdf

Dimitrov, D.V. (2016). Medical Internet of Things and Big Data in healthcare. *Healthcare Informatics Research*, 22(3), 156-163. https://doi.org/10.4258/hir.2016.22.3.156

Dixit, A.K. and Nalebuff, B.J. (2008). *The Art of Strategy: A Game Theorist's Guide to Success in Business and Life*. New York, NY: W.W. Norton & Company, Inc.

Dogac, A., Yuksel, M., Ertürkmen, G., Kabak, Y., Namli, T. and Yıldız, M. et al. (2014). Healthcare information technology infrastructures in Turkey. *IMIA Yearbook*, 9(1), 228-234. http://dx.doi.org/10.15265/iy-2014-0001

Dolan, P.L. (2010). EMRs missing the mark on care coordination efforts. *American Medical News*. Retrieved from http://www.ama-assn.org/amednews/2010/01/18/bica0118.htm

Dorri, A., Kanhere, S.S. and Jurdak, R. (2016). Blockchain in Internet of Things: Challenges and solutions. *ArXiv:1608.05187 [Cs]*. http://arxiv.org/abs/1608.05187

Dorri, A., Kanhere, S.S. and Jurdak, R. (2017). Towards an optimized blockchain for IoT. *2nd International Conference on Internet-of-Things Design and Implementation (IoTDI)*, 173-178. https://doi.org/10.1145/3054977.3055003

Dorri, A., Kanhere, S.S., Jurdak, R. and Gauravaram, P. (2017). Blockchain for IoT security and privacy: The case study of a smart home. *Pervasive Computing and Communications Workshops (PerCom Workshops)*, 618-623. https://doi.org/10.1109/PERCOMW.2017.7917634

Doukas, C. and Maglogiannis, I. (2012). Bringing IoT and cloud computing towards pervasive healthcare. *2012 Sixth International Conference on Innovative Mobile and Internet Services in Ubiquitous Computing*, 922-926. https://doi.org/10.1109/IMIS.2012.26

Du, L. and Lu, W. (2016). U.S. Health-care system ranks as one of the least-efficient. Retrieved from https://www.bloomberg.com/news/articles/2016-11-8/u-s-health-care-system-ranks-as-one-of-the-least-efficient

References

Dyer, K.A. and Thompson, C.D. (2001). Medical Internet ethics: A field in evolution. pp. 1287-1291. *In*: V.L. Patel, R. Rogers and R. Haux (Eds.), MedInfo 2001. Amsterdam: IOS Press.

Ebel, T., George, K., Larsen, E., Neal, E., Shah, K. and Shi, D. (2012). Strength in unity: The promise of global standards in healthcare. Retrieved from http://www.mckinsey.com/search.aspx?q=Strength+in+Unity

Eggbeer, W. (2015). Launching a Provider Led Health Plan?. BDC. Retrieved 17 March 2016, from http://www.bdcadvisors.com/launching-a-provider-led-health-plan/

Eisenhardt, K. (1989). Building theories from case study research. *Academy of Management Review*, 14(4), 532-550. http://dx.doi.org/10.5465/amr.1989.4308385

Eisenmann, T.R. (2008). Managing Proprietary and Shared Platforms, *California Management Review*, 50(4), 31-53.

El Sawy, O., Pereira, F. and Fife, E. (2008). The VISOR framework: Business model definition for new marketspaces in the networked digital industry. *In*: R.S. Sharma, M. Tan and F. Pereira (Eds.), Understanding the IDM Marketplace. Singapore: IGI Publishing.

Emergency Monitoring and Prevention | EMERGE Project | FP6 | CORDIS | European Commission. (2007, February 1). https://cordis.europa.eu/project/rcn/80505/factsheet/en

Engelhardt, M.A. (2017). Hitching healthcare to the chain: An introduction to blockchain technology in the healthcare sector. *Technology Innovation Management Review*, 7(10), 22-34. https://doi.org/10.22215/timreview/1111

Etemadi, M., Inan, O.T., Heller, J.A., Hersek, S., Klein, L. and Roy, S. (2016). A wearable patch to enable long-term monitoring of environmental, activity and hemodynamics variables. *IEEE Transactions on Biomedical Circuits and Systems*, 10(2), 280-288. https://doi.org/10.1109/TBCAS.2015.2405480

Ethereum Classic. (n.d.). Retrieved October 25, 2019, from https://ethereumclassic.org/

Eysenbach, G. (2001). What is e-health? *Journal of Medical Internet Research*, 3(2), 20.

Fact Sheet: President Obama's Precision Medicine Initiative. (2015, January 30). Whitehouse.Gov. https://obamawhitehouse.archives.gov/the-press-office/2015/01/30/fact-sheet-president-obama-s-precision-medicine-initiative

Fan, K., Wang, S., Ren, Y., Li, H. and Yang, Y. (2018). MedBlock: Efficient and secure medical data sharing via blockchain. *Medical Systems*, 42(8), 136. https://doi.org/10.1007/s10916-018-0993-7

Fang, F., Aabith, S., Homer-Vanniasinkam, S. and Tiwari, M.K. (2017). High-resolution 3D printing for healthcare underpinned by small-scale fluidics. *3D Printing in Medicine*, 167-206. Elsevier. https://doi.org/10.1016/B978-0-08-100717-4.00023-5

Fang, R., Pouyanfar, S., Yang, Y., Chen, S.-C. and Iyengar, S.S. (2016). Computational health informatics in the Big Data age: A survey. *ACM Computing Surveys (CSUR)*, 49(1), 1-36. https://doi.org/10.1145/2932707

Feero, W.G., Wicklund, C.A. and Veenstra, D. (2018). Precision medicine, genome sequencing, and improved population health. *American Medical Association*, 319(19), 1979. https://doi.org/10.1001/jama.2018.2925

Fernández-Caramés, T.M. and Fraga-Lamas, P. (2018). A review on the use of blockchain for the Internet of Things. *IEEE Access*, 6, 32979-33001. https://doi.org/10.1109/ACCESS.2018.2842685

Ferrag, M.A., Derdour, M., Mukherjee, M., Derhab, A., Maglaras, L. and Janicke, H. (2019). Blockchain technologies for the Internet of Things: Research issues and challenges. *IEEE Internet of Things*, 6(2), 2188-2204. https://doi.org/10.1109/JIOT.2018.2882794

Ferre, F., de Belvis, A.G., Valerio, L., Longhi, S., Lazzari, A., Fattore, G., . . . Maresso, A. (2014). Italy: Health system review. *Health Syst Transit*, 16(4), 1-168.

Firdaus, A., Anuar, N.B., Razak, M.F.A., Hashem, I.A.T., Bachok, S. and Sangaiah, A.K. (2018). Root exploit detection and features optimization: Mobile device and blockchain based medical data management. *Medical Systems*, 42(6), 112. https://doi.org/10.1007/s10916-018-0966-x

Fisher, E., McClellan, M. and Safran, D. (2011). Building the path to accountable care. *New England Journal of Medicine*, 365(26), 2445-2447. http://dx.doi.org/10.1056/nejmp1112442

Fitbit Charge™ Wireless Activity + Sleep Wristband. (n.d.). Retrieved October 20, 2019, from https://www.fitbit.com/uk/charge

Fitness Tracker | Fitness Bands | Activity Tracker | Garmin. (n.d.). Retrieved October 20, 2019, from https://buy.garmin.com/en-NZ/NZ/cIntoSports-c571-p1.html

Fleming, N.S., Culler, S.D., McCorkle, R., Becker, E.R. and Ballard, D.J. (2011). The Financial and nonfinancial costs of implementing electronic health records in primary care practices. *Health Affairs*, 30(3), 481-489.

Fonseca, M., Cardoso, H., Ferreira, N., Loureiro, T., Gomes, I. and Quaresma, C. (2019). VR4Neuropain: Interactive rehabilitation system. *12th International Joint Conference on Biomedical Engineering Systems and Technologies (BIOSTEC)*, 285-290. Scopus.

Ford, E., Hesse, B. and Huerta, T. (2016). Personal health record use in the United States: Forecasting future adoption levels. *Journal of Medical Internet Research*, 18(3), e73. http://dx.doi.org/10.2196/jmir.4973

Ford, E., Wells, R. and Bailey, B. (2004). Sustainable network advantages. *Health Care Management Review*, 29(2), 159-169. http://dx.doi.org/10.1097/00004010-200404000-00009

Foundation Medicine. (2015, November 3). Foundation Medicine Announces 2015 Third Quarter Results and Recent Highlights. Foundation Medicine, Inc. http://investors.foundationmedicine.com/news-releases/news-release-details/foundation-medicine-announces-2015-third-quarter-results-and

Frakt, A. (2015). Accountable Care Organizations: Like H.M.O.s, but Different. Nytimes.com. Retrieved from http://www.nytimes.com/2015/01/20/upshot/accountable-care-organizations-like-hmos-but-different.html?_r=0

Frakt, A. and Mayes, R. (2012). Beyond capitation: How new payment experiments seek to find the 'Sweet Spot' in amount of risk providers and payers bear. *Health Affairs*, 31(9), 1951-1958. http://dx.doi.org/10.1377/hlthaff.2012.0344

Fuchs, V. (2009). Health reform: Getting the essentials right. *Health Affairs*, 28(2), w180-w183. http://dx.doi.org/10.1377/hlthaff.28.2.w180

Fuchs, V. (2014). Critiquing US Health Care. *JAMA*, 312(20), 2095. http://dx.doi.org/10.1001/jama.2014.14114

Gajanayake, R., Sahama, T. and Iannella, R. (2013). E-health in Australia and elsewhere: A comparison and lessons for the near future. pp. 26-32. *In*: H. Grain and L. Schaper (Eds.), Health Informatics: Digital Health Service Delivery - The Future is now! Amsterdam: IOS Press BV.

Garets, D. and Davis, M. (2005). Electronic patient records. *Healthcare Informatics Online*, 4.

Genomics-Enabled Learning Health Care Systems: Gathering and Using Genomic Information to Improve Patient Care and Research: Workshop Summary. (2015). National Academies Press (US). http://www.ncbi.nlm.nih.gov/books/NBK294179/

References

Gill, P., Stewart, K., Treasure, E. and Chadwick, B. (2008). Methods of data collection in qualitative research: Interviews and focus groups. *BDJ*, 204(6), 291-295. http://dx.doi.org/10.1038/bdj.2008.192

Giuffrida, J.P., Lerner, A., Steiner, R. and Daly, J. (2008). Upper-extremity stroke therapy task discrimination using motion sensors and electromyography. *IEEE Transactions on Neural Systems and Rehabilitation Engineering*, 16(1), 82-90. https://doi.org/10.1109/TNSRE.2007.914454

Glaser, J. (2007). Financial incentives for healthcare IT adoption. *Healthcare Financial Management*, 61(4), 118-120.

Global Health Observatory. (2016). Country statistics. World Health Organization. Retrieved 12 January 2017, from http://www.who.int/gho/countries/en/

Golab, M.R., Breedon, P.J. and Vloeberghs, M. (2016). A wearable headset for monitoring electromyography responses within spinal surgery. *European Spine Journal*, 25(10), 3214-3219. https://doi.org/10.1007/s00586-016-4626-x

Goldman, D. (2009, July 2). Electronic health records: A checkup. CNNMoney.com. Retrieved from http://money.cnn.com/2009/07/02/news/economy/stimulus_electronic_health_records

Goldsmith, J. (2012). Health reform: A political tragedy. *Health Affairs*, 31(1), 245-246.

Goldstein, J. (2009, April). Big challenges await health-records transition. *The Wall Street Journal*. Retrieved from http://online.wsj.com/article/SB124027664223937475.html

Grandori, A. (Ed.). (2012). *Interfirm Networks: Organization and Industrial Competitiveness*. Routledge.

Grandori, A. and Neri, M. (1999). The fairness properties of interfirm networks. pp. 41-66. *In*: A. Grandori (Ed.), Interfirm Networks: Organization and Industrial Competitiveness. London: Routledge.

Grandy, C. (2006). Through a glass darkly: An economic view of fairness, globalization, and states. pp. 49-60. *In*: J.A. Dator, R. Pratt and Y. Seo (Eds.), Fairness, Globalization, and Public Institutions: East Asia and Beyond. USA: University of Hawaii Press.

Griggs, K.N., Ossipova, O., Kohlios, C.P., Baccarini, A.N., Howson, E.A. and Hayajneh, T. (2018). Healthcare blockchain system using smart contracts for secure automated remote patient monitoring. *Medical Systems*, 42(7), 130. https://doi.org/10.1007/s10916-018-0982-x

Gubbins, E. (2009). How Telcos could conquer the cloud. ConnectedPlanetOnline.com. Retrieved from http://connectedplanetonline.com/business_services/news/cloud-computing-telecom-service-providers-0401/

Guest, J. and Quincy, L. (2013). Consumers gaining ground in health care. *JAMA*, 310(18), 1939. http://dx.doi.org/10.1001/jama.2013.282000

Guidi, A., Salvi, S., Ottaviano, M., Gentili, C., Bertschy, G., de Rossi, D., Scilingo, E. and Vanello, N. (2015). Smartphone application for the analysis of prosodic features in running speech with a focus on bipolar disorders: System performance evaluation and case study. *Sensors*, 15(11), 28070-28087. https://doi.org/10.3390/s151128070

Gummesson, E. (1991). *Qualitative Methods in Management Research* (Revised Edition). London: Sage.

Gunapal, P., Kannapiran, P., Teow, K., Zhu, Z., Xiaobin You, A. and Saxena, N. et al. (2016). Setting up a regional health system database for seamless population health management in Singapore. *Proceedings of Singapore Healthcare*, 25(1), 27-34. http://dx.doi.org/10.1177/2010105815611440

Guo, R., Shi, H., Zhao, Q. and Zheng, D. (2018). Secure attribute-based signature scheme with multiple authorities for blockchain in electronic health records systems. *IEEE Access*, 6, 11676-11686. https://doi.org/10.1109/ACCESS.2018.2801266

Guo, W., Xiong, N., Vasilakos, A.V., Chen, G. and Cheng, H. (2011). Multi-source temporal data aggregation in wireless sensor networks. *Wireless Personal Communications*, 56(3), 359-370. https://doi.org/10.1007/s11277-010-9976-9

Gupta, M. (2017). *Blockchain For Dummies® IBM Limited Edition*. 51.

Hafen, E., Kossmann, D. and Brand, A. (2014). Health Data Cooperatives – Citizen Empowerment. *Methods Inf Med*, 53(2), 82-86. http://dx.doi.org/10.3414/me13-02-0051

Hagel III, J. and Rayport, J.F. (1997). The Coming Battle for Customer Information. Harvard Business Review, 75(1), 53-65.

Hagglund, M. and Koch, S. (2015). Commentary: Sweden rolls out online access to medical records and is developing new e-health services to enable people to manage their care. *BMJ*, 350(feb11 2), h359-h359. http://dx.doi.org/10.1136/bmj.h359

Hall, M., Frank, E., Holmes, G., Pfahringer, B., Reutemann, P. and Witten, I.H. (2009). The WEKA data mining software: An update. *SIGKDD Explor. Newsl.*, 11(1), 10-18. https://doi.org/10.1145/1656274.1656278

Hallfors, N.G., Alhawari, M., Abi Jaoude, M., Kifle, Y., Saleh, H., Liao, K., Ismail, M. and Isakovic, A.F. (2018). Graphene oxide: Nylon ECG sensors for wearable IoT healthcare—nanomaterial and SoC interface. *Analog Integrated Circuits and Signal Processing*, 96(2), 253-260. https://doi.org/10.1007/s10470-018-1116-6

Hammond, W.E., Bailey, C., Boucher, P., Spohr, M. and Whitaker, P. (2010). Connecting information to improve health. *Health Affairs*, 29(2), 284-288.

Handelman, G.S., Kok, H.K., Chandra, R.V., Razavi, A.H., Lee, M.J. and Asadi, H. (2018). eDoctor: Machine learning and the future of medicine. *Internal Medicine*, 284(6), 603-619. https://doi.org/10.1111/joim.12822

Hardjono, T. and Pentland, A. (2019). Verifiable Anonymous Identities and Access Control in Permissioned Blockchains. *ArXiv, abs/1903.04584.*

He, K., Ge, D. and He, M. (2017). Big Data analytics for genomic medicine. *Molecular Sciences*, 18(2), 412. https://doi.org/10.3390/ijms18020412

HealthIT.gov. (2013). Meaningful Use Criteria and How to Attain Meaningful Use of EHRs. Retrieved 7 March 2016, from https://www.healthit.gov/providers-professionals/how-attain-meaningful-use

Herzlinger, R. (2010). Healthcare reform and its implications for the U.S. economy. *Business Horizons*, 53(2), 105-117. http://dx.doi.org/10.1016/j.bushor.2009.10.002

Hickey, K.T., Bakken, S., Byrne, M.W., Bailey, D. (Chip) E., Demiris, G., Docherty, S.L., Dorsey, S.G., Guthrie, B.J., Heitkemper, M.M., Jacelon, C.S., Kelechi, T.J., Moore, S.M., Redeker, N.S., Renn, C.L., Resnick, B., Starkweather, A., Thompson, H., Ward, T.M., McCloskey, D. J., … Grady, P.A. (2019). Precision health: Advancing symptom and self-management science. *Nursing Outlook*. https://doi.org/10.1016/j.outlook.2019.01.003

Hijazi, S., Page, A., Kantarci, B. and Soyata, T. (2016). Machine learning in cardiac health monitoring and decision support. *Computer*, 49(11), 38-48. Scopus. https://doi.org/10.1109/MC.2016.339

Hill, J.W. and Powell, P. (2009). The national healthcare crisis: Is eHealth a key solution? *Business Horizons*, 52(3), 265-277.

Hill, J.W., Langvardt, A.W. and Massey, A.P. (2007). Law, information technology, and medical errors: Toward a national healthcare information network approach

References

to improving patient care and reducing medical malpractice costs. *Journal of Law, Technology & Policy*, 2, 159-238.

Hölbl, M., Kompara, M., Kamišalić, A. and Nemec Zlatolas, L. (2018). A systematic review of the use of blockchain in healthcare. *Symmetry*, 10(10), 470. https://doi.org/10.3390/sym10100470

Home | Ethereum. (n.d.). Ethereum.Org. Retrieved October 25, 2019, from https://ethereum.org

Home—Qtum. (n.d.). Retrieved October 25, 2019, from https://qtum.org/en

Hooshmand, M., Zordan, D., Del Testa, D., Grisan, E. and Rossi, M. (2017). Boosting the battery life of wearables for health monitoring through the compression of biosignals. *IEEE Internet of Things*, 4(5), 1647-1662. https://doi.org/10.1109/JIOT.2017.2689164

Hossain, M.S. and Muhammad, G. (2016). Healthcare Big Data voice pathology assessment framework. *IEEE Access*, 4, 7806-7815. https://doi.org/10.1109/ACCESS.2016.2626316

Hossain, M.S., Rahman, Md. A. and Muhammad, G. (2017). Towards energy-aware cloud-oriented cyber-physical therapy system. *Future Generation Computer Systems*. https://doi.org/10.1016/j.future.2017.08.045

Hou, H. (2017). The application of blockchain technology in e-government in China. *26th International Conference on Computer Communication and Networks (ICCCN)*, 1-4. https://doi.org/10.1109/ICCCN.2017.8038519

Hsiao, K.-F. and Rashvand, H.F. (2015). Data modeling mobile augmented reality: Integrated mind and body rehabilitation. *Multimedia Tools and Applications*, 74(10), 3543-3560. https://doi.org/10.1007/s11042-013-1649-8

Huang, Q., Jing, S., Yi, J. and Zhen, W. (2016). *Innovative Testing and Measurement Solutions for Smart Grid* (1st ed.). Wiley-IEEE Press.

Huang, Y., Zheng, H., Nugent, C., McCullagh, P., Black, N., Hawley, M. and Mountain, G. (2011). Knowledge discovery from lifestyle profiles to support self-management of Chronic Heart Failure. *Computing in Cardiology*, 397-400. IEEE Conference, Hangzhou, China.

HUAWEI Band 3 Pro, built-in GPS sport band, smart wearable HUAWEI Global. (n.d.). Retrieved October 20, 2019, from https://consumer.huawei.com/en/wearables/band3-pro/

Hussein, Ahmed F., Arun Kumar, N., Ramirez-Gonzalez, G., Abdulhay, E., Tavares, J.M.R.S. and de Albuquerque, V.H.C. (2018). A medical records managing and securing blockchain based system supported by a Genetic Algorithm and Discrete Wavelet Transform. *Cognitive Systems Research*, 52, 1-11. https://doi.org/10.1016/j.cogsys.2018.05.004

Hussein, Ahmed Faeq, Kumar, N.A., Burbano-Fernandez, M., Ramírez-González, G., Abdulhay, E. and De Albuquerque, V.H.C. (2018). An automated remote cloud-based heart rate variability monitoring system. *IEEE Access*, 6, 77055-77064. https://doi.org/10.1109/ACCESS.2018.2831209

Hwang, J. and Christensen, C. (2008). Disruptive Innovation In Health Care Delivery: A Framework For Business-Model Innovation. *Health Affairs*, 27(5), 1329-1335. http://dx.doi.org/10.1377/hlthaff.27.5.1329

Ichikawa, D., Kashiyama, M. and Ueno, T. (2017). Tamper-resistant mobile health using blockchain technology. *MHealth and UHealth*, 5(7), e111. https://doi.org/10.2196/mhealth.7938

Ilie, V., Van Slyke, C., Parikh, M. and Courtney, J. (2009). Paper versus electronic medical records: The effects of access on physicians' decisions to use complex information technologies. *Decision Sciences*, 40(2), 213-241. http://dx.doi.org/10.1111/j.1540-5915.2009.00227.x

Illumina Introduces the HiSeq X™ Ten Sequencing System. (2014, January 14). https://www.businesswire.com/news/home/20140114006291/en/Illumina-Introduces-HiSeq-X%E2%84%A2-Ten-Sequencing-System

Institute of Policy Studies (2012). IPS Study: Singapore could be 'extremely aged' by 2050. Retrieved from http://lkyspp.nus.edu.sg/ips/wp-content/uploads/sites/2/2013/06/Yahoo_IPS-Study-Singapore-could-be-extremely-aged-by-2050_030512.pdf

Jacob, J. (2013). Financial incentives are spurring growth of electronic health records. *BMJ*, 347(aug09 1), f4901-f4901. http://dx.doi.org/10.1136/bmj.f4901

Jain, P., Agarwal, A., Behara, R. and Baechle, C. (2019). HPCC based framework for COPD readmission risk analysis. *Big Data*, 6(1). Scopus. https://doi.org/10.1186/s40537-019-0189-0

Jarosławski, S. and Saberwal, G. (2014). In eHealth in India today, the nature of work, the challenges and the finances: An interview-based study. *BMC Medical Informatics and Decision Making*, 14(1). http://dx.doi.org/10.1186/1472-6947-14-1

Ji, Y., Zhang, J., Ma, J., Yang, C. and Yao, X. (2018). BMPLS: Blockchain-based multi-level privacy-preserving location sharing scheme for telecare medical information systems. *Medical Systems*, 42(8), 147. https://doi.org/10.1007/s10916-018-0998-2

Jiang, S., Cao, J., Wu, H., Yang, Y., Ma, M. and He, J. (2018). BlocHIE: A Blockchain-based platform for healthcare information exchange. *IEEE International Conference on Smart Computing (SMARTCOMP)*, 49-56. https://doi.org/10.1109/SMARTCOMP.2018.00073

Jin, H., Luo, Y., Li, P. and Mathew, J. (2019). A review of secure and privacy-preserving medical data sharing. *IEEE Access*, 7, 61656-61669. https://doi.org/10.1109/ACCESS.2019.2916503

Jo, B.W., Khan, R.M.A. and Lee, Y.-S. (2018). Hybrid blockchain and Internet-of-Things network for underground structure health monitoring. *Sensors*, 18(12), 4268. https://doi.org/10.3390/s18124268

Jones, D.K. (2012). The fate of health care reform - What to expect in 2012. *New England Journal of Medicine*, 366(4), e7.

Joshi, A.P., Han, M. and Wang, Y. (2018). A survey on security and privacy issues of blockchain technology. *Mathematical Foundations of Computing*, 1(2), 121. https://doi.org/10.3934/mfc.2018007

Joslyn, J.S. (2001). Healthcare E-Commerce: Connecting with Patients. *Journal of Healthcare Information Management*, 15(1), 73-84.

Joslyn, J.S. (2001). Healthcare e-Commerce: Connecting with patients. *Journal of Healthcare Information Management*, 15(1), 73-84.

Jovanovic, B. and Rousseau, P.L. (2005). Chapter 18—General purpose technologies. pp. 1181-1224. *In*: P. Aghion and S.N. Durlauf (Eds.), Handbook of Economic Growth (Vol. 1). Elsevier. https://doi.org/10.1016/S1574-0684(05)01018-X

Juneja, A. and Marefat, M.M. (2018). Leveraging blockchain for retraining deep learning architecture in patient-specific arrhythmia classification. *IEEE EMBS International Conference on Biomedical and Health Informatics*, 393-397. https://doi.org/10.1109/BHI.2018.8333451

Kaelber, D. and Pan, E.C. (2008). The value of personal health record (PHR) systems. *AMIA Annual Symposium Proceedings*, Vol. 2008, p. 343. American Medical Informatics Association.

Kafhali, S.E. and Salah, K. (2018). Performance modelling and analysis of Internet of Things enabled healthcare monitoring systems. *IET Networks*, 8(1), 48-58. https://doi.org/10.1049/iet-net.2018.5067

Kandiah, G. and Gossain, S. (1998). Reinventing value: The new business ecosystem. *Strategy & Leadership*, 26(5), 28-33. doi:10.1108/eb054622

Kantarjian, H. and Yu, P.P. (2015). Artificial intelligence, big data, and cancer. *Oncology*, 1(5), 573-574. https://doi.org/10.1001/jamaoncol.2015.1203

Kasthurirathne, S.N., Mamlin, B., Kumara, H., Grieve, G. and Biondich, P. (2015). Enabling better interoperability for healthcare: Lessons in developing a standards based application programing interface for electronic medical record systems. *Medical Systems*, 39(11), 182. https://doi.org/10.1007/s10916-015-0356-6

Kaur, H., Alam, M.A., Jameel, R., Mourya, A.K. and Chang, V. (2018). A proposed solution and future direction for blockchain-based heterogeneous medicare data in cloud environment. *Medical Systems*, 42(8), 156. https://doi.org/10.1007/s10916-018-1007-5

Kayyali, B., Knott, D. and Van Kuiken, S. (2013). The 'big data' revolution in healthcare: Accelerating value and innovation. Retrieved from http://www.mckinsey.com/insights/health_systems_and_services/the_big-data_revolution_in_us_health_care

Kelleher, J. (2015). OpenGov speaks to Mr James Chia, Integrated Health Information Systems, Singapore. Opengovasia.com. Retrieved 24 February 2016, from http://www.opengovasia.com/articles/6611-opengov-speaks-to-mr-james-chia-integrated-health-information-systems

Kellermann, A.L. and Jones, S.S. (2013). What it will take to achieve the as-yet-unfulfilled promises of health information technology. *Health Affairs*, 32(1), 63-68.

Kellogg, R.A., Dunn, J. and Snyder, M.P. (2018). Personal Omics for Precision Health. *Circulation Research*, 122(9), 1169-1171. https://doi.org/10.1161/CIRCRESAHA.117.310909

Kelly, G. and Muers, S. (2002). *Creating Public Value: An Analytical Framework for Public Service Reform*. Strategy Unit, Cabinet Office, UK.

Ketchersid, T. (2013). Big Data in nephrology: Friend or foe? *Blood Purification*, 36(3-4), 160-164. https://doi.org/10.1159/000356751

Khalik, S. (2015). National University Hospital recalls 178 children for TB tests. *The Straits Times*. Retrieved from http://www.straitstimes.com/singapore/health/national-university-hospital-recalls-178-children-for-tb-tests

Khan, M. (2014). 14 predictions for healthcare in 2014. *Health Management Technology*, 13. Retrieved from http://www4.healthmgttech.com/articles/201401/14-predictions-for-healthcare-in-2014.php

Khan, N., Yaqoob, I., Hashem, I.A.T., Inayat, Z., Ali, W.K.M., Alam, M., Shiraz, M. and Gani, A. (2014). Big Data: Survey, technologies, opportunities, and challenges. *The Scientific World*, 2014, 712826. https://doi.org/10.1155/2014/712826

Kheirkhahan, M., Nair, S., Davoudi, A., Rashidi, P., Wanigatunga, A.A., Corbett, D.B., Mendoza, T., Manini, T.M. and Ranka, S. (2019). A smartwatch-based framework for real-time and online assessment and mobility monitoring. *Biomedical Informatics*, 89, 29-40. https://doi.org/10.1016/j.jbi.2018.11.003

Khezr, S., Moniruzzaman, M., Yassine, A. and Benlamri, R. (2019). Blockchain Technology in healthcare: A comprehensive review and directions for future research. *Applied Sciences*, 9(9), 1736. https://doi.org/10.3390/app9091736

Khojasteh, S.B., Villar, J.R., Chira, C., González, V.M. and de la Cal, E. (2018). Improving fall detection using an on-wrist wearable accelerometer. *Sensors* (Basel, Switzerland), 18(5). https://doi.org/10.3390/s18051350

Kilgore, C. (1999). Patients take the wheel with Internet health records. *Telehealth Magazine*, 5(7), 7-8.

Kim, A.H., Kendrick, D.E., Moorehead, P.A., Nagavalli, A., Miller, C.P., Liu, N.T., Wang, J.C. and Kashyap, V.S. (2016). Endovascular aneurysm repair simulation can lead to decreased fluoroscopy time and accurately delineate the proximal seal zone. *Vascular Surgery*, 64(1), 251-258. https://doi.org/10.1016/j.jvs.2016.01.050

Kinsman, L., Rotter, T., James, E., Snow, P. and Willis, J. (2010). What is a clinical pathway? Development of a definition to inform the debate. *BMC Medicine*, 8, 31. https://doi.org/10.1186/1741-7015-8-31

Kivits, J. (2013). e-Health and renewed sociological approaches to health and illness. pp. 213-226. *In*: K. Orton-Johnson and N. Prior (Eds.), Digital Sociology: Critical Perspectives. London: Palgrave Macmillan UK.

Klonoff, D.C. (2013). Twelve modern digital technologies that are transforming decision making for diabetes and all areas of health care. *Diabetes Science and Technology*, 7(2), 291-295. https://doi.org/10.1177/193229681300700201

Kno.e.sis | From Information to Meaning. (n.d.). Retrieved October 29, 2019, from http://knoesis.org/

Koh, A. and Cheah, J. (2015). No hospital is an island-emerging new roles of the acute general hospital in the Singapore healthcare ecosystem. *Future Hospital Journal*, 2(2), 121-124.

Kohout, J. and Kukačka, M. (2014). Real-time modelling of fibrous muscle. *Computer Graphics Forum*, 33(8), 1-15. Scopus. https://doi.org/10.1111/cgf.12354

Konrad, W. and Peter, H. (2007). Balancing of Benefits and Disadvantages using IT-Integration to Support the Health Care value-added Chain. pp. 101-114. *In*: A. Hein, W. Thoben, H. Appelrath and P. Jensch (Eds.), European Conference on eHealth 2007, Proceedings of the ECEH'07 held in Oldenburg, Germany, 11-12 October 2007. Bonn, GI: Gesellschaft für Informatik.

Korn, O., Buchweitz, L., Rees, A., Bieber, G., Werner, C. and Hauer, K. (2019). Using augmented reality and gamification to empower rehabilitation activities and elderly persons: A study applying design thinking. pp. 219-229. *In*: T.Z. Ahram (Ed.), Advances in Artificial Intelligence, Software and Systems Engineering. Springer International Publishing. https://doi.org/10.1007/978-3-319-94229-2_21

Kosba, A., Miller, A., Shi, E., Wen, Z. and Papamanthou, C. (2016). Hawk: The blockchain model of cryptography and privacy-preserving smart contracts. *2016 IEEE Symposium on Security and Privacy (SP)*, 839-858. https://doi.org/10.1109/SP.2016.55

Krittanawong, C., Bomback, A.S., Baber, U., Bangalore, S., Messerli, F.H. and Wilson Tang, W.H. (2018). Future direction for using artificial intelligence to predict and manage hypertension. *Current Hypertension Reports*, 20(9), 75. https://doi.org/10.1007/s11906-018-0875-x

Krumholz, H.M., Terry, S.F. and Waldstreicher, J. (2016). Data acquisition, curation, and use for a continuously learning health system. *JAMA*, 316(16), 1669-1670. doi:10.1001/jama.2016.12537

Kuo, A.M.H. (2011). Opportunities and challenges of cloud computing to improve health care services. *Journal of Medical Internet Research*, 13(3).

References

Kuo, T.-T. and Ohno-Machado, L. (2018). ModelChain: Decentralized privacy-preserving healthcare predictive modeling framework on private blockchain networks. *ArXiv:1802.01746 [Cs]*. http://arxiv.org/abs/1802.01746

Kuo, T.-T., Kim, H.-E. and Ohno-Machado, L. (2017). Blockchain distributed ledger technologies for biomedical and health care applications. *American Medical Informatics Association*, 24(6), 1211-1220. https://doi.org/10.1093/jamia/ocx068

Kvale, S. and Brinkmann, S. (2009). *InterViews: Learning the Craft of Qualitative Research Interviewing*. Los Angeles: Sage Publications.

Kwapisz, J.R., Weiss, G.M. and Moore, S.A. (2011). Activity recognition using cell phone accelerometers. *SIGKDD Explor. Newsl.*, 12(2), 74-82. https://doi.org/10.1145/1964897.1964918

LaMontagne, C. (2013). NerdWallet Health finds Medical Bankruptcy accounts for majority of personal bankruptcies – NerdWallet. *NerdWallet*. Retrieved from http://www.nerdwallet.com/blog/health/managing-medical-bills/nerdwallet-health-study-estimates-56-million-americans-65-struggle-medical-bills-2013/

Landman, J.H. (2015). Narrow networks and healthcare competition. *Healthcare Financial Management*, 69(4), 102-104.

Laney, D. (2001, June 2). 3D data management: Controlling data volume, velocity and variety. https://blogs.gartner.com/doug-laney/files/2012/01/ad949-3D-Data-Management-Controlling-Data-Volume-Velocity-and-Variety.pdf

Lassetter, J.K. (2010, January). 2010: Change & Progress: Forecast 2010: Hospitals: HIEs to Transform. *Health Management Technology*, 18-18. Retrieved from http://www4.healthmgttech.com/ebook/201001/resources/index.htm.

Lawrence, D.R., Palacios-González, C. and Harris, J. (2016). Artificial intelligence: The shylock syndrome. *Cambridge Quarterly of Healthcare Ethics*, 25(2), 250-261. https://doi.org/10.1017/S0963180115000559

Lee, B.-G., Lee, B.-L. and Chung, W.-Y. (2014). Mobile healthcare for automatic driving sleep-onset detection using wavelet-based EEG and respiration signals. *Sensors* (Basel, Switzerland), 14(10), 17915-17936. https://doi.org/10.3390/s141017915

Lee, C.K., Kim, Y., Lee, N., Kim, B., Kim, D. and Yi, S. (2017). Feasibility study of utilization of action camera, GoPro Hero 4, Google Glass, and Panasonic HX-A100 in Spine Surgery. *Spine*, 42(4), 275-280. https://doi.org/10.1097/BRS.0000000000001719

Lee, J. (2015). Singapore's Health Cloud Edged Out Hundreds of Global Submissions to Snag Prestigious DataCloud Enterprise Cloud Award in Monaco. Ihis.com.sg. Retrieved from https://www.ihis.com.sg/MediaCentre/mr/Pages/Singapore%E2%80%99s-Health-Cloud-Edged-Out-Hundreds-of-Global-Submissions-to-Snag-Prestigious-DataCloud-Enterprise-Cloud-Award-in-M.aspx

Lee, Y. and Lee, C.H. (2018). Augmented reality for personalized nanomedicines. *Biotechnology Advances*, 36(1), 335-343. https://doi.org/10.1016/j.biotechadv.2017.12.008

Leedy, P. and Ormrod, J. (2010). *Practical Research: Planning and Design* (9th ed.). Upper Saddle River, NJ: Merrill Prentice Hall. Thousand Oaks: Sage Publications.

Leslie, H. (2005). Commentary: The patient's memory stick may complement electronic health records. *Australian Health Review*, 29(4), 401-405.

Lewis, N. (2010). Microsoft Aims to Alleviate Health IT Cloud Concerns. *Information Week*. Retrieved from http://www.informationweek.com/news/healthcare/security-privacy/showArticle.jhtml?articleID=225702414

Li, H., Zhu, L., Shen, M., Gao, F., Tao, X. and Liu, S. (2018). Blockchain-based data preservation system for medical data. *Medical Systems*, 42(8), 141. https://doi.org/10.1007/s10916-018-0997-3

Li, Xiaohuan, Huang, X., Li, C., Yu, R. and Shu, L. (2019). EdgeCare: Leveraging edge computing for collaborative data management in mobile healthcare systems. *IEEE Access*, 7, 22011-22025. https://doi.org/10.1109/ACCESS.2019.2898265

Li, Xiaoqi, Jiang, P., Chen, T., Luo, X. and Wen, Q. (2017). A survey on the security of blockchain systems. *Future Generation Computer Systems*. https://doi.org/10.1016/j.future.2017.08.020

Liang, X., Zhao, J., Shetty, S., Liu, J. and Li, D. (2017). Integrating blockchain for data sharing and collaboration in mobile healthcare applications. *28th Annual International Symposium on Personal, Indoor, and Mobile Radio Communications (PIMRC)*, 1-5. https://doi.org/10.1109/PIMRC.2017.8292361

Liao, L., Fox, D. and Kautz, H.A. (2005). Location-based activity recognition using relational markov networks. *19th International Joint Conference on Artificial Intelligence (IJCAI)*, 773-778.

Liao, L., Patterson, D.J., Fox, D. and Kautz, H. (2007). Learning and inferring transportation routines. *Artificial Intelligence*, 171(5), 311-331. https://doi.org/10.1016/j.artint.2007.01.006

Lim, M.K. (2004). Quest for quality care and patient safety: The case of Singapore. *Quality and Safety in Health Care*, 13(1), 71-75.

Lim, M.K. (2005). Transforming Singapore health care: Public-private partnership. *Annals-Academy of Medicine Singapore*, 34(7), 461.

Lim, S.S., Allen, K., Bhutta, Z.A., Dandona, L., Forouzanfar, M.H., Fullman, N., . . . Murray, C.J.L. (2016). Measuring the health-related sustainable development goals in 188 countries: A baseline analysis from the Global Burden of Disease Study 2015. *The Lancet*, 388(10053), 1813-1850. doi:10.1016/S0140-6736(16)31467-2

Lin, C., Xiong, N., Park, J.H. and Kim, T. (2009). Dynamic power management in new architecture of wireless sensor networks. *International Journal of Communication Systems*, 22(6), 671-693. https://doi.org/10.1002/dac.989

Lin, I.-C. and Liao, T.-C. (2017). A survey of blockchain security issues and challenges. *Network Security*, 19(5), 653-659. https://doi.org/10.6633/IJNS.201709.19(5).01

Litecoin—Open source P2P digital currency. (n.d.). Retrieved October 25, 2019, from https://litecoin.org/

Liu, Y. and Tang, P. (2018). The prospect for the application of the surgical navigation system based on artificial intelligence and augmented reality. *International Conference on Artificial Intelligence and Virtual Reality (AIVR)*, 244-246. https://doi.org/10.1109/AIVR.2018.00056

Lohr, S. (2007, June). Who Pays for Efficiency? *The New York Times*. Retrieved from http://www.nytimes.com/2007/06/11/business/businessspecial3/11save.html?_r=1&ref=businessspecial3

Lopez-Perez, L., Hernandez, L., Ottaviano, M., Martinelli, E., Poli, T., Licitra, L., Arredondo, M.T. and Fico, G. (2019). BD2Decide: Big Data and models for personalized head and neck cancer decision support. *IEEE Symposium on Computer-Based Medical Systems*, 2019-June, 67-68. Scopus. https://doi.org/10.1109/CBMS.2019.00024

Loria, K. (2016, March 4). Therapists have created a virtual "heroin cave" in an attempt to help addicts. *Business Insider*. https://www.businessinsider.com/university-of-houston-heroin-cave-virtual-reality-2016-3

References

Lowe, S.A. and Ólaighin, G. (2014). Monitoring human health behaviour in one's living environment: A technological review. *Medical Engineering & Physics*, 36(2), 147-168. https://doi.org/10.1016/j.medengphy.2013.11.010

Luce, R. and Raiffa, H. (1957). *Games and Decisions*. New York: Wiley.

Luciano, C.J., Banerjee, P.P., Sorenson, J.M., Foley, K.T., Ansari, S.A., Rizzi, S., Germanwala, A.V., Kranzler, L., Chittiboina, P. and Roitberg, B.Z. (2013). Percutaneous spinal fixation simulation with virtual reality and haptics. *Neurosurgery*, 72(1), 89-96. https://doi.org/10.1227/NEU.0b013e3182750a8d

Luštrek, M. and Kaluža, B. (2009). Fall detection and activity recognition with machine learning. *Informatica*, 33(2). http://www.informatica.si/index.php/informatica/article/view/238

Lv, Z., Esteve, C., Chirivella, J. and Gagliardo, P. (2017). Serious game based personalized healthcare system for dysphonia rehabilitation. *Pervasive and Mobile Computing*, 41, 504-519. Scopus. https://doi.org/10.1016/j.pmcj.2017.04.006

Lyles, C.R., Lunn, M.R., Obedin-Maliver, J. and Bibbins-Domingo, K. (2018). The new era of precision population health: Insights for the All of Us Research Program and beyond. *Translational Medicine*, 16(1), 211. https://doi.org/10.1186/s12967-018-1585-5

Magyar, G., Balsa, J., Cláudio, A.P., Carmo, M.B., Neves, P., Alves, P., Félix, I.B., Pimenta, N. and Guerreiro, M.P. (2019). Anthropomorphic virtual assistant to support self-care of type 2 diabetes in older people: A perspective on the role of artificial intelligence. *14th International Joint Conference on Computer Vision, Imaging and Computer Graphics Theory and Applications (VISIGRAPP)*, 1, 323-331. Scopus.

Mahmud, S., Iqbal, R. and Doctor, F. (2016). Cloud enabled data analytics and visualization framework for health-shocks prediction. *Future Generation Computer Systems*, 65, 169-181. https://doi.org/10.1016/j.future.2015.10.014

Mamoshina, P., Ojomoko, L., Yanovich, Y., Ostrovski, A., Botezatu, A., Prikhodko, P., Izumchenko, E., Aliper, A., Romantsov, K., Zhebrak, A., Ogu, I.O. and Zhavoronkov, A. (2018). Converging blockchain and next-generation artificial intelligence technologies to decentralize and accelerate biomedical research and healthcare. *Oncotarget*, 9(5), 5665-5690. https://doi.org/10.18632/oncotarget.22345

Mangan, D. (2015). US health care: Spending a lot, getting the least. *CNBC*. Retrieved 16 March 2016, from http://www.cnbc.com/2015/10/08/us-health-care-spending-is-high-results-arenot-so-good.html

Martinez, D., Feijoo, F., Zayas-Castro, J., Levin, S. and Das, T. (2016). A strategic gaming model for health information exchange markets. *Health Care Management Science*. http://dx.doi.org/10.1007/s10729-016-9382-2

Mavrogiorgou, A., Kiourtis, A., Perakis, K., Pitsios, S. and Kyriazis, D. (2019). IoT in healthcare: Achieving interoperability of high-quality data acquired by IoT medical devices. *Sensors*, 19(9), 1978. https://doi.org/10.3390/s19091978

McCann, E. (2014). HIE goes live with eHealth Exchange. *Healthcare IT News*. Retrieved 17 March 2016, from http://www.healthcareitnews.com/news/hie-goes-live-ehealth-exchange

McCarthy, M. (2015). US healthcare spending will reach 20% of GDP by 2024, says report. *BMJ*, h4204. http://dx.doi.org/10.1136/bmj.h4204

McFarlane, C., Beer, M., Brown, J. and Prendergast, N. (2017). Patientory: A Healthcare Peer-to-Peer EMR Storage Network v1.0. 19.

McKesson Corporation. (2007). EMR Return on Investment: Improving Efficiency and Quality with an Electronic Medical Record. [White Paper].

McLean, T. (2007). Commoditization of the international teleradiology market. *Journal of Health Services Research & Policy*, 12(2), 120-122. http://dx.doi.org/10.1258/135581907780279503

Melillo, P., Orrico, A., Scala, P., Crispino, F. and Pecchia, L. (2015). Cloud-based smart health monitoring system for automatic cardiovascular and fall risk assessment in hypertensive patients. *Medical Systems*, 39(10), 109. https://doi.org/10.1007/s10916-015-0294-3

Menachemi, N. and Collum, T. (2011). Benefits and drawbacks of electronic health record systems. *Risk Management and Healthcare Policy*, 4, 47-55. doi:10.2147/RMHP.S12985

Meng-Kin, L. (1998). Health care systems in transition II. Singapore, Part I. An overview of health care systems in Singapore. *Journal of Public Health Medicine*, 20, 16-22.

Mensink, W. and Birrer, F.A. (2010). The role of expectations in system innovation: The electronic health record, immoderate goal or achievable necessity. *Central European Journal of Public Policy*, 4(1), 36-59.

Mesko, B. (2017). The role of artificial intelligence in precision medicine. *Expert Review of Precision Medicine and Drug Development*, 2(5), 239-241. https://doi.org/10.1080/23808993.2017.1380516

Mettler, M. (2016). Blockchain technology in healthcare: The revolution starts here. *2016 IEEE 18th International Conference on E-Health Networking, Applications and Services (Healthcom)*, 1-3. https://doi.org/10.1109/HealthCom.2016.7749510

Mettler, T. and Eurich, M. (2012). A "design-pattern"-based approach for analyzing e-health business models. *Health Policy and Technology*, 1(2), 77-85.

Mettler, T., Rohner, P. and Baacke, L. (2008). Improving data quality of health information systems: a holistic design-oriented approach. *Proceedings of the European Conference on Information Systems ECIS '08*, pp. 1883-1893.

Michel, S. (2014). Capture more value. *Harvard Business Review*, 92(10), 78-85.

Middleton, B. (2008). EHRs, PHRs, and HIE: Impact on patient safety, healthcare quality and costs. Presented at the seminar entitled Patient-Centred Computing and eHealth: Transforming Healthcare Quality, Boston, MA.

Miles, M. and Huberman, A. (1994). *Qualitative Data Analysis an Expanded Sourcebook* (1st ed.). Thousand Oaks: Sage.

Miller, R.H., West, C., Brown, T.M., Sim, I. and Ganchoff, C. (2005). The value of electronic health records in solo or small group practices. *Health Affairs*, 24(5), 1127-1137.

MOH Press Release. (2015). One-stop health information and services portal provides access to key medical records and health content | Ministry of Health. Moh.gov.sg. Retrieved 20 February 2016, from https://www.moh.gov.sg/content/moh_web/home/pressRoom/pressRoomItemRelease/2015/one-stop-health-information-and-services-portal-provides-access-.html

Moisseiev, E. and Mannis, M.J. (2016). Evaluation of a portable artificial vision device among patients with low vision. *JAMA Ophthalmology*, 134(7), 748-752. https://doi.org/10.1001/jamaophthalmol.2016.1000

Moncada-Torres, A., Leuenberger, K., Gonzenbach, R., Luft, A. and Gassert, R. (2014). Activity classification based on inertial and barometric pressure sensors at different anatomical locations. *Physiological Measurement*, 35(7), 1245-1263. https://doi.org/10.1088/0967-3334/35/7/1245

Monero: Titles.themoneroproject. (n.d.). Getmonero.Org, The Monero Project. Retrieved October 25, 2019, from https://getmonero.org/the-monero-project/index.html

Montgomery, K. (2016). Pay-for-Performance Systems Reward Doctors and Providers for Efficiency. About.com Health. Retrieved from http://healthinsurance.about.com/od/faqs/f/p4p.htm

Moore, J.F. (1996). *The Death of Competition: Leadership and Strategy in the Age of Business Ecosystems*. HarperBusiness.

Morales-Arroyo, M. and Sharma, R.S. (2009). Deriving value in digital media networks. *International Journal of Computer Science and Security*, 3(2), 126-137.

Moses III, H., Matheson, D., Dorsey, E., George, B., Sadoff, D. and Yoshimura, S. (2013). The anatomy of health care in the United States. *JAMA*, 310(18), 1947. http://dx.doi.org/10.1001/jama.2013.281425

MOSKUS – MObile musculoSKeletal User Self-management. (2015, December 3). Norsk Regnesentral. https://www.nr.no/nb/projects/moskus-%E2%80%93-mobile-musculoskeletal-user-self-management

Mossialos, E., Wenzl, M., Osborn, R. and Sarnak, D. (2016). 2015 International Profiles of Health Care Systems. The Commonwealth Fund. Retrieved from http://www.commonwealthfund.org/publications/fund-reports/2016/jan/international-profiles-2015

Mullin, R. (2016). EHR Incentive Program Status Report January 2016. HITECH Answers: Meaningful Use, HER, HIPAA News. Retrieved from http://www.hitechanswers.net/ehr-incentive-program-status-report-january-2016/

Muttitt, S. (2011, March). National Electronic Health Record Perspectives: Singapore. Interview, Virtua.

Muttitt, S., McKinnon, S. and Rainey, S. (2012). *Singapore's National Electronic Health Record (NEHR): The Journey to 2012 and Beyond*. Presentation, Singapore.

MyCode Community Health Initiative. (n.d.). Retrieved November 5, 2019, from https://www.geisinger.org/mycode

Myerson, R. (1991). *Game Theory*. Cambridge, Mass.: Harvard University Press.

Nakamoto, S. (2008). *Bitcoin: A Peer-to-Peer Electronic Cash System* (p. 9).

Naranjo-Hernández, D., Roa, L.M., Reina-Tosina, J. and Estudillo-Valderrama, M.Á. (2012). SoM: A smart sensor for human activity monitoring and assisted healthy ageing. *IEEE Transactions on Bio-Medical Engineering*, 59(11), 3177-3184. https://doi.org/10.1109/TBME.2012.2206384

Narayanan, H.A.J. and Güneş, M.H. (2011, January). Ensuring access control in cloud provisioned healthcare systems. pp. 247-251. *In: Consumer Communications and Networking Conference (CCNC), 2011 IEEE*.

Nausheen, F. and Begum, S.H. (2018). Healthcare IoT: Benefits, vulnerabilities and solutions. *2nd International Conference on Inventive Systems and Control (ICISC)*, 517-522. https://doi.org/10.1109/ICISC.2018.8399126

Neo-project. (n.d.). *Neo Smart Economy*. Retrieved October 25, 2019, from https://neo.org/

Neupert, P. (2009). Re-inventing Healthcare. *Health Management Technology*, 30(3), 8-10.

Ng, A.T.S., Sy, C. and Li, J. (2011, December). A system dynamics model of Singapore healthcare affordability. pp. 1-13. *In: Simulation Conference (WSC), Proceedings of the 2011 Winter*. IEEE.

Ng, K. (2013). Private Cloud to Cut Costs for Singapore's Public Health. Ihis.com.sg. Retrieved from https://www.ihis.com.sg/MediaCentre/NewsArticles/Pages/Private-Cloud-To-Cut-Costs-For-Singapore's-Public-Health.aspx

Nigam, V.K. and Bhatia, S. (2016). Impact of cloud computing on health care. *International Research Journal of Engineering and Technology*, 3(5), 2804-2810.

Nikoloudakis, Y., Pallis, E., Mastorakis, G., Mavromoustakis, C.X., Skianis, C. and Markakis, E.K. (2019). Vulnerability assessment as a service for fog-centric ICT ecosystems: A healthcare use case. *Peer-to-Peer Networking and Applications*, 12(5), 1216-1224. https://doi.org/10.1007/s12083-019-0716-y

Nilsson, N.J. (1980). *Principles of Artificial Intelligence*. Morgan Kaufmann Publishers Inc.

Nilsson, Nils J. (1998). *Artificial Intelligence: A New Synthesis*. Morgan Kaufmann Publishers Inc.

Nonnemann, L., Haescher, M., Aehnelt, M., Bieber, G., Diener, H. and Urban, B. (2019). Health@Hand A Visual Interface for eHealth Monitoring. 2019-June. Scopus. https://doi.org/10.1109/ISCC47284.2019.8969647

Noor, K.B.M. (2008). Case study: A strategic research methodology. *American Journal of Applied Sciences*, 5(11), 1602-1604.

Nurjono, M., Valentijn, P.P., Bautista, M.A.C., Lim, Y.W. and Vrijhoef, H.J. (2016). A prospective validation study of a rainbow model of integrated care measurement tool in Singapore. *International Journal of Integrated Care*, 16(1).

O'Malley, A.S., Grossman, J.M., Cohen, G.R., Kemper, N.M. and Hoangmai. (2009). Are electronic medical records helpful for care coordination? Experiences of physician practices. *Journal of General Internal Medicine*, doi:10.1007/s11606-009-1195-2

Obamacare Facts. (2016). ObamaCare | Health Insurance Exchange. Retrieved from http://obamacarefacts.com/obamacare-health-insurance-exchange/

Oderkirk, J., Ronchi, E. and Klazinga, N. (2013). International comparisons of health system performance among OECD countries: Opportunities and data privacy protection challenges. *Health Policy*, 112(1-2), 9-18. http://dx.doi.org/10.1016/j.healthpol.2013.06.006

Öksüz, N., Shcherbatyi, I., Kowatsch, T. and Maass, W. (2018). A data-analytical system to predict therapy success for obese children. *39th International Conference on Information Systems (ICIS)*, 1-17. Scopus. https://cocoa.ethz.ch/downloads/2018/10/None_Oksuz%20et%20al%202018%20Data-analytical%20System%20Obese%20Children.pdf

Olivera, P., Danese, S., Jay, N., Natoli, G. and Peyrin-Biroulet, L. (2019). Big Data in IBD: A look into the future. *Nature Reviews Gastroenterology & Hepatology*, 16(5), 312-321. https://doi.org/10.1038/s41575-019-0102-5

Ølnes, S., Ubacht, J. and Janssen, M. (2017). Blockchain in government: Benefits and implications of distributed ledger technology for information sharing. *Government Information Quarterly*, 34(3), 355-364. https://doi.org/10.1016/j.giq.2017.09.007

Omar, A.A., Bhuiyan, M.Z.A., Basu, A., Kiyomoto, S. and Rahman, M.S. (2019). Privacy-friendly platform for healthcare data in cloud based on blockchain environment. *Future Generation Computer Systems*, 95, 511-521. https://doi.org/10.1016/j.future.2018.12.044

Orlikoff, J.E. and Totten, M.K. (2010). Preparing for the impact of healthcare reform. Focusing on patient-access processes is key to success. *Healthcare Executive*, 25(1), 66-67.

Osborne, M. and Rubinstein, A. (1994). *A Course in Game Theory*. Cambridge, Mass. [u.a.]: MIT Press.

Otoum, S., Kantarci, B. and Mouftah, H. (2018). Adaptively supervised and intrusion-aware data aggregation for wireless sensor clusters in critical infrastructures. *IEEE International Conference on Communications (ICC)*, 1-6. https://doi.org/10.1109/ICC.2018.8422401

References

Otoum, S., Kantarci, B. and Mouftah, H.T. (2019). On the feasibility of deep learning in sensor network intrusion detection. *IEEE Networking Letters*, 1(2), 68-71. https://doi.org/10.1109/LNET.2019.2901792

Pagliari, C., Sloan, D., Gregor, P., Sullivan, F., Detmer, D. and Kahan, J. et al. (2005). What is eHealth (4): A scoping exercise to map the field. *J. Med. Internet Res*, 7(1), e9.

Palanica, A., Docktor, M.J., Lee, A. and Fossat, Y. (2019). Using mobile virtual reality to enhance medical comprehension and satisfaction in patients and their families. *Perspectives on Medical Education*, 8(2), 123-127. Scopus. https://doi.org/10.1007/s40037-019-0504-7

Panesar, A. (2019). *Machine Learning and AI for Healthcare: Big Data for Improved Health Outcomes*. Apress. https://doi.org/10.1007/978-1-4842-3799-1

Pappas, I.P., Popovic, M.R., Keller, T., Dietz, V. and Morari, M. (2001). A reliable gait phase detection system. *IEEE Transactions on Neural Systems and Rehabilitation Engineering*, 9(2), 113-125. https://doi.org/10.1109/7333.928571

Paraskevopoulos, I. and Tsekleves, E. (2013). Use of gaming sensors and customised exergames for Parkinson's disease rehabilitation: A proposed virtual reality framework. *5th International Conference on Games and Virtual Worlds for Serious Applications (VS-GAMES)*, 1-5. https://doi.org/10.1109/VS-GAMES.2013.6624247

Parente, S.T. (2000). Beyond the hype: A taxonomy of e-health business models. *Health Affairs*, 19(6), 89-102.

Parmar, R., Mackenzie, I., Cohn, D. and Gann, D. (2014). The new patterns of innovation. *Harvard Business Review*, 92(1/2), 86-95.

Patel, V. (2019). A framework for secure and decentralized sharing of medical imaging data via blockchain consensus. *Health Informatics*, 25(4), 1398-1411. https://doi.org/10.1177/1460458218769699

Patton, M. (1987). *How to Use Qualitative Methods in Evaluation* (1st ed.). Newbury Park, Calif.: Sage Publications.

Patton, M. (1990). *Qualitative Evaluation and Research Methods* (1st ed.). Newbury Park, Calif.: Sage Publications.

Paun, I., Sauciuc, D., Iosif, N., Stan, O., Perşe, A., Dehelean, C. and Miclea, L. (2011). Local EHR management based on openEHR and EN13606. *J Med Syst*, 35(4), 585-590. http://dx.doi.org/10.1007/s10916-009-9395-1

Payton, F.C. (2003). e-Health models leading to business-to-employee commerce in the human resources function. *Journal of Organizational Computing and Electronic Commerce*, 13(2), 147-161.

Peters, K., Niebling, M., Slimmer, C., Green, T., Webb, J.M. and Schumacher, R. (2009, February). Usability guidance for improving the user interface and adoption of online personal health records. *In*: Proc the Human Factors and Ergonomics Society Annual Meeting, Oakbrook Terrace, IL, US: User Centric, Inc (Vol. 11, pp. 704-708).

Peters, K., Niebling, M., Slimmer, C., Green, T., Webb, J.M. and Schumacher, R. (2009). Usability Guidance for Improving the User Interface and Adoption of Online Personal Health Records. UserCentric.com. Retrieved September 15, 2009, from http://www.scribd.com/doc/21811231/Usability-Guidance-for-Improving-the-User-Interface-and-Adoption-of-Online-Personal-Health-Records

Peterson, K.J., Deeduvanu, R., Kanjamala, P. and Mayo, K. (2016). A Blockchain-Based Approach to Health Information Exchange Networks.

Pierleoni, P., Belli, A., Maurizi, L., Palma, L., Pernini, L., Paniccia, M. and Valenti, S. (2016). A wearable fall detector for elderly people based on AHRS and barometric sensor. *IEEE Sensors*, 16(17), 6733-6744. https://doi.org/10.1109/JSEN.2016.2585667

Pierleoni, P., Belli, A., Palma, L., Pellegrini, M., Pernini, L. and Valenti, S. (2015). A high reliability wearable device for elderly fall detection. *IEEE Sensors*, 15, 4544-4553. https://doi.org/10.1109/jsen.2015.2423562

Poon, J. and Dryja, T. (n.d.). The Bitcoin Lightning Network: Scalable Off-Chain Instant Payments. 59.

Porter, M. (2009). A strategy for health care reform—Toward a value-based system. *New England Journal of Medicine*, 361(2), 109-112. http://dx.doi.org/10.1056/nejmp0904131

Porter, M. (2010). What is value in health care? *New England Journal of Medicine*, 363(26), 2477-2481. http://dx.doi.org/10.1056/nejmp1011024

Powell, P.T. and Laufer, R. (2010). The promises and constraints of consumer-directed healthcare. *Business Horizons*, 53(2), 171-182.

Pramanik, M.I., Lau, R.Y.K., Demirkan, H. and Azad, Md. A.K. (2017). Smart health: Big Data enabled health paradigm within smart cities. *Expert Systems with Applications*, 87, 370-383. https://doi.org/10.1016/j.eswa.2017.06.027

Provider-Led Health Insurers - Healthcare Consumer Satisfaction - Summary - Accenture. (2007). Accenture.com. Retrieved 4 November 2009, from https://www.accenture.com/us-en/insight-healthcare-consumers-one-stop-shopping-summary.aspx

Puppala, M., He, T., Chen, S., Ogunti, R., Yu, X., Li, F., Jackson, R. and Wong, S.T.C. (2015). METEOR: An enterprise health informatics environment to support evidence-based medicine. *IEEE Transactions on Bio-Medical Engineering*, 62(12), 2776-2786. https://doi.org/10.1109/TBME.2015.2450181

Purcarea, O. (2009, October 23). European Health Telematics Association (EHTEL) Symposium – A decade of dedicated support to eHealth in Europe. Microsoft EMEA Health Blog. Retrieved November 05, 2009 from http://blogs.msdn.com/ms_emea_health_blog/archive/2009/10/23/ehtel-a-decade-of-dedicated-support-to-ehealth-in-europe.aspx

Qi, J., Chen, L., Leister, W. and Yang, S. (2015). Towards knowledge driven decision support for personalized home-based self-management of chronic diseases. *12th International Conference on Ubiquitous Intelligence and Computing and 12th International Conference on Autonomic and Trusted Computing and 15th International Conference on Scalable Computing and Communications and Associated Workshops (UIC-ATC-ScalCom)*, 1724-1729. https://doi.org/10.1109/UIC-ATC-ScalCom-CBDCom-IoP.2015.313

Qi, J., Yang, P., Fan, D. and Deng, Z. (2015). A survey of physical activity monitoring and assessment using Internet of Things technology. *IEEE International Conference on Computer and Information Technology; Ubiquitous Computing and Communications; Dependable, Autonomic and Secure Computing; Pervasive Intelligence and Computing*, 2353-2358. https://doi.org/10.1109/CIT/IUCC/DASC/PICOM.2015.348

Qi, J., Yang, P., Min, G., Amft, O., Dong, F. and Xu, L. (2017). Advanced Internet of Things for personalised healthcare systems: A survey. *Pervasive and Mobile Computing*, 41, 132-149. https://doi.org/10.1016/j.pmcj.2017.06.018

Qiu, Y. and Ma, M. (2018). Secure Group Mobility Support for 6LoWPAN Networks. *IEEE Internet of Things*, 5(2), 1131-1141. https://doi.org/10.1109/JIOT.2018.2805696

Qu, Z., Lau, C.W., Nguyen, Q.V., Zhou, Y. and Catchpoole, D.R. (2019). Visual analytics of genomic and cancer data: A systematic review. *Cancer Informatics*, 18, 117693511983554. https://doi.org/10.1177/1176935119835546

References

Raghupathi, W. and Kesh, S. (2009). Designing electronic health records versus total digital health systems: A systemic analysis. *Systems Research and Behavioral Science*, 26(1), 63-79. doi: 10.1002/sres.918.

Rahman, Md. A., Hassanain, E., Rashid, Md. M., Barnes, S.J. and Hossain, M.S. (2018). Spatial blockchain-based secure mass screening framework for children with dyslexia. *IEEE Access*, 6, 61876-61885. https://doi.org/10.1109/ACCESS.2018.2875242

Rajasekar, S., Philominathan, P. and Chinnathambi, V. (2006), "Research methodology", available at: http://arxiv.org/abs/physics/0601009 (accessed 28 March 2014).

Ramesh, M., Wu, X. and Howlett, M. (2015). Second best governance? Governments and governance in the imperfect world of health care delivery in China, India and Thailand in Comparative Perspective. *Journal of Comparative Policy Analysis: Research and Practice*, 17(4), 342-358. http://dx.doi.org/10.1080/13876988.2014.889903

Rapaport, A. and Chammah, A. (1965). *Prisoner's Dilemma*. Ann Arbor: Univ. of Michigan Press.

Rappaport, A. (1992, May-June). CFOs and strategists: Forging a common framework. *Harvard Business Review*, 70, 84-91.

Rau, J. (2014). Medicare Fines 2,610 Hospitals in Third Round of Readmission Penalties. *Kaiser Health News*. Retrieved from http://khn.org/news/medicare-readmissions-penalties-2015/

Reddy, S., Mun, M., Burke, J., Estrin, D., Hansen, M. and Srivastava, M. (2010). Using mobile phones to determine transportation modes. *ACM Transactions on Sensor Networks (TOSN)*, 6(2), 13. https://doi.org/10.1145/1689239.1689243

Reed, G. (2007). Barriers to Successful EHR Implementation. EHRScope.com. Retrieved 15 March 2014, from http://www.ehrscope.com

Regan, D.B., Pusatli, O.T., Lutton, E. and Athauda, D.R. (2009, June). Securing an EHR in a health sector digital ecosystem. *In: 2009 3rd IEEE International Conference on Digital Ecosystems and Technologies*.

Regional Health Systems (RHS). (n.d.). Retrieved February 24, 2015, from http://www.aic.sg/page.aspx?id=137

RFA-MD-15-013: NIMHD Transdisciplinary Collaborative Centers for Health Disparities Research Focused on Precision Medicine (U54). (1992, April). https://grants.nih.gov/grants/guide/rfa-files/RFA-MD-15-013.html

Roberta. (2016). EHR Incentive Program Status Report January 2016. HITECH Answers: Meaningful Use, HER. HIPAA News. Retrieved 5 March 2016, from http://www.hitechanswers.net/ehr-incentive-program-status-report-january-2016/

Robey, D., Boudreau, M. and Rose, G. (2000). Information technology and organizational learning: A review and assessment of research. *Accounting, Management and Information Technologies*, 10(2), 125-155. http://dx.doi.org/10.1016/s0959-8022(99)00017-x

Roehrs, A., da Costa, C.A. and da Rosa Righi, R. (2017). OmniPHR: A distributed architecture model to integrate personal health records. *Biomedical Informatics*, 71, 70-81. https://doi.org/10.1016/j.jbi.2017.05.012

Roman-Belmonte, J.M., De la Corte-Rodriguez, H. and Rodriguez-Merchan, E.C. (2018). How blockchain technology can change medicine. *Postgraduate Medicine*, 130(4), 420-427. https://doi.org/10.1080/00325481.2018.1472996

Romano, D. and Schmid, G. (2017). Beyond bitcoin: A critical look at blockchain-based systems. *Cryptography*, 1(2), 15. https://doi.org/10.3390/cryptography1020015

Romeo, J. (2013). Connect the dots. *Journal of AHIMA/American Health Information Management Association*, 84(3), 22-26.

Romero, L.E., Chatterjee, P. and Armentano, R.L. (2016). An IoT approach for integration of computational intelligence and wearable sensors for Parkinson's disease, diagnosis and monitoring. *Health and Technology*, 6(3), 167-172. https://doi.org/10.1007/s12553-016-0148-0

Rosenberg, L.B. (1992). The Use of Virtual Fixtures as Perceptual Overlays to Enhance Operator Performance in Remote Environments. https://apps.dtic.mil/docs/citations/ADA292450

Rosenthal, E. (2013). Health Care's Road to Ruin. Nytimes.com. Retrieved from http://www.nytimes.com/2013/12/22/sunday-review/health-cares-road-to-ruin.html

Ross, J., Stevenson, F., Lau, R. and Murray, E. (2015). Exploring the challenges of implementing e-health: A protocol for an update of a systematic review of reviews. *BMJ Open*, 5(4), http://dx.doi.org/10.1136/bmjopen-2014-006773

Rothenhaus, T. (2015). Using data to increase patient engagement in health care. *Harvard Business Review*. Retrieved from https://hbr.org/2015/06/using-data-to-increase-patient-engagement-in-health-care

Rouse, M. (2014). What is HITECH Act (Health Information Technology for Economic and Clinical Health Act)? - Definition from WhatIs.com. *SearchHealthIT*. Retrieved from http://searchhealthit.techtarget.com/definition/HITECH-Act

Rudin, R.S., Jones, S.S., Shekelle, P., Hillestad, R.J. and Keeler, E.B. (2014). The value of health information technology: Filling the knowledge gap. *The American Journal of Managed Care*, 20(SP17), eSP1-eSP8.

Russell, S. and Norvig, P. (2009). *Artificial Intelligence: A Modern Approach* (3rd ed.). Pearson.

Sackett, D.L., Rosenberg, W.M.C., Gray, J.A.M., Haynes, R.B. and Richardson, W.S. (1996). Evidence based medicine: What it is and what it isn't: It's about integrating individual clinical expertise and the best external evidence. *BMJ: British Medical Journal*, 312(7023), 71-72.

Sahoo, P.K., Mohapatra, S.K. and Wu, S.-L. (2018). SLA based healthcare Big Data analysis and computing in cloud network. *Journal of Parallel and Distributed Computing*, 119, 121-135. https://doi.org/10.1016/j.jpdc.2018.04.006

Saia, R. (2018). Internet of Entities (IoE): A blockchain-based distributed paradigm to security. *ArXiv:1808.08809 [Cs]*. http://arxiv.org/abs/1808.08809

Saia, R., Carta, S., Recupero, D. and Fenu, G. (2019). Internet of Entities (IoE): A blockchain-based distributed paradigm for data exchange between wireless-based devices. *8th International Conference on Sensor Networks*, 77-84. https://doi.org/10.5220/0007379600770084

Sakr, S. and Elgammal, A. (2016). Towards a comprehensive data analytics framework for smart healthcare services. *Big Data Research*, 4, 44-58. https://doi.org/10.1016/j.bdr.2016.05.002

Salarian, A., Russmann, H., Wider, C., Burkhard, P.R., Vingerhoets, F.J.G. and Aminian, K. (2007). Quantification of tremor and bradykinesia in Parkinson's disease using a novel ambulatory monitoring system. *IEEE Transactions on Bio-Medical Engineering*, 54(2), 313-322. https://doi.org/10.1109/TBME.2006.886670

Saldana, J. (2009). An introduction to codes and coding. *The Coding Manual for Qualitative Researchers*. 1-31.

Saleem, J.J., Flanagan, M.E., Wilck, N.R., Demetriades, J. and Doebbeling, B.N. (2013). The next-generation electronic health record: Perspectives of key leaders from the US Department of Veterans Affairs. *Journal of the American Medical Informatics Association*, 20(e1), e175-e177.

References

Santos, D.F.S., Almeida, H.O. and Perkusich, A. (2015). A personal connected health system for the Internet of Things based on the constrained application protocol. *Computers & Electrical Engineering*, 44, 122-136. https://doi.org/10.1016/j.compeleceng.2015.02.020

Saratzis, A., Calderbank, T., Sidloff, D., Bown, M.J. and Davies, R.S. (2017). Role of simulation in Endovascular Aneurysm Repair (EVAR) training: A preliminary study. *Vascular and Endovascular Surgery*, 53(2), 193-198. https://doi.org/10.1016/j.ejvs.2016.11.016

Sardini, E., Serpelloni, M. and Pasqui, V. (2015). Wireless wearable T-shirt for posture monitoring during rehabilitation exercises. *IEEE Transactions on Instrumentation and Measurement*, 64(2), 439-448. https://doi.org/10.1109/TIM.2014.2343411

Scerri, A. and James, P. (2010). Accounting for sustainability: Combining qualitative and quantitative research in developing 'indicators' of sustainability. *International Journal of Social Research Methodology*, 13(1), 41-53.

Schaefer, C. (2011). C-A3-04: The Kaiser Permanente Research Program on genes, environment and health: A resource for genetic epidemiology in adult health and aging. *Clinical Medicine & Research*, 9(3-4), 177-178. https://doi.org/10.3121/cmr.2011.1020.c-a3-04

SearchHealthIT. (2010). State HIE list from the HITECH Act. Retrieved 7 March 2016, from http://searchhealthit.techtarget.com/tip/State-HIE-list-from-the-HITECH-Act

Seidman, I. (2013). *Interviewing as Qualitative Research: A Guide for Researchers in Education and the Social Sciences*. Teachers College Press.

Semenov, I., Kopanitsa, G., Denisov, D., Alexandr, Y., Osenev, R. and Andreychuk, Y. (2018). Patients decision aid system based on FHIR profiles. *Medical Systems*, 42(9), 166. https://doi.org/10.1007/s10916-018-1016-4

Serbanati, L., Ricci, F., Mercurio, G. and Vasilateanu, A. (2011). Steps towards a digital health ecosystem. *Journal of Biomedical Informatics*, 44(4), 621-636. http://dx.doi.org/10.1016/j.jbi.2011.02.011

Shaban-Nejad, A., Lavigne, M., Okhmatovskaia, A. and Buckeridge, D.L. (2017). PopHR: A knowledge-based platform to support integration, analysis, and visualization of population health data. *New York Academy of Sciences*, 1387(1), 44-53. https://doi.org/10.1111/nyas.13271

Shae, Z. and Tsai, J. (2018). Transform blockchain into distributed parallel computing architecture for precision medicine. *38th International Conference on Distributed Computing Systems (ICDCS)*, 1290-1299. https://doi.org/10.1109/ICDCS.2018.00129

Shae, Z. and Tsai, J.J.P. (2017). On the design of a blockchain platform for clinical trial and precision medicine. *37th International Conference on Distributed Computing Systems (ICDCS)*, 1972-1980. https://doi.org/10.1109/ICDCS.2017.61

Shafer, S.M., Smith, H.J. and Linder, J.C. (2005). The power of business models. *Business Horizons*, 48(3), 199-207.

Sharfstein, J., Fontanarosa, P. and Bauchner, H. (2013). Critical Issues in US Health Care. *JAMA*, 310(18), 1945. http://dx.doi.org/10.1001/jama.2013.282124

Sharma, R.S. and Kshetri, N. (2020). Digital healthcare: Historical development, applications, and future research directions. Guest Editors' Introduction to Special Issue on Digital Health. *International Journal of Information Management*. https://doi.org/10.1016/j.ijinfomgt.2020.102105

Sharma, R.S. (2011). Towards a strategic tool for the business modeling of IDM startups - Implementing the ADVISOR framework. *International Journal of Applied Engineering and Technology*, 1(1), 68-79, October-December.

Sharma, R.S. and Bhattacharya, S. (2013). Knowledge dilemmas within organizations: Resolutions from game theory. *Knowledge-Based Systems*, 45, 100-113. http://dx.doi.org/10.1016/j.knosys.2013.02.011

Sharma, R.S., Pereira, F., Ramasubbu, N., Tan, M. and Tschang, T. (2010). Assessing value creation and value capture in digital business ecosystems. *International Journal of Information Technology*, 16(2).

Shay, R. (2016). EHR Adoption Rates in 2014 and 2015 | Practice Fusion. Practice Fusion Blog. Retrieved from http://www.practicefusion.com/blog/ehr-adoption-rates/

Shen, B., Guo, J. and Yang, Y. (2019). MedChain: Efficient healthcare data sharing via blockchain. *Applied Sciences*, 9(6), 1207. https://doi.org/10.3390/app9061207

Sheth, A., Jaimini, U. and Yip, H.Y. (2018). How will the Internet of Things enable augmented personalized health? *IEEE Intelligent Systems*, 33(1), 89-97. https://doi.org/10.1109/MIS.2018.012001556

Shi, L. and Singh, D. (2015). *Major Characteristics of US health care delivery. Essentials of the US Health Care System.* Sudbury, MA: Jones and Bartlett Publishers, 1-25.

Shimrat, O. (2009). Cloud Computing and Healthcare: Bad Weather or Sunny Forecast? San Diego Physician.org, 26-29.

Shmanske, S. (1996). Information Asymmetries in Health Services. *The Independent Review*, 1(2), 191-200.

Showell, C. and Nohr, C. (2012). How Should We Define eHealth, and Does the Definition Matter? Retrieved from http://ebooks.iospress.nl/volume/quality-of-life-through-quality-of-information

Shu, L., Zhang, Y., Yu, Z., Yang, L.T., Hauswirth, M. and Xiong, N. (2010). Context-aware cross-layer optimized video streaming in wireless multimedia sensor networks. *Supercomputing*, 54(1), 94-121. https://doi.org/10.1007/s11227-009-0321-6

Shuaib, Saleous, Shuaib and Zaki. (2019). Blockchains for secure digitized medicine. *Personalized Medicine*, 9(3), 35. https://doi.org/10.3390/jpm9030035

Shum, E. and Lee, C.E. (2014). Population-based healthcare: The experience of a regional health system. *Ann Acad Med Singapore*, 43(12), 564-565.

Singapore Business Review. (2016). Public hospitals may struggle to provide sufficient capacity for patients by 2030. Retrieved 7 July 2016, from http://sbr.com.sg/healthcare/news/singapore%E2%80%99s-public-hospitals-may-struggle-provide-sufficient-capacity-patients-203-0

Singapore Parliamentary Report. (2014). Hospital readmission rates | Ministry of Health. Moh.gov.sg. Retrieved 26 February 2016, from https://www.moh.gov.sg/content/moh_web/home/pressRoom/Parliamentary_QA/2014/hospital-readmission-rates.html

Singapore Parliamentary Report. (2014). MOH 2012 Committee of Supply Speech Healthcare 2020: Improving Accessibility, Quality and Affordability for Tomorrow's Challenges | Ministry of Health. Moh.gov.sg. Retrieved 26 February 2016, from https://www.moh.gov.sg/content/moh_web/home/pressRoom/speeches_d/2012/moh_2012_committeeofsupplyspeechhealthcare2020improvingaccessibi.html

Singapore Parliamentary Report. (2015). Future phase of NEHR | Ministry of Health. Moh.gov.sg. Retrieved 26 February 2016, from https://www.moh.gov.sg/content/moh_web/home/pressRoom/Parliamentary_QA/2015/future-phase-of-nehr.html

Sinha, P., Sunder, G., Bendale, P., Mantri, M. and Dande, A. (2013). *Electronic Health Record: Standards, Coding Systems, Frameworks, and Infrastructures* (1st ed.). John Wiley & Sons, Inc.

Sinha, R.S., Wei, Y. and Hwang, S.-H. (2017). A survey on LPWA technology: LoRa and NB-IoT. *ICT Express*, 3(1), 14-21. https://doi.org/10.1016/j.icte.2017.03.004

References

Sinnapolu, G. and Alawneh, S. (2018). Integrating wearables with cloud-based communication for health monitoring and emergency assistance. *Internet of Things*, 1-2, 40-54. https://doi.org/10.1016/j.iot.2018.08.004

Sittig, D.F. and Singh, H. (2012). Electronic Health Records and National Patient-Safety Goals. *New England Journal of Medicine*, 367(19), 1854-1860.

Smith, J. and Medalia, C. (2015). Health Insurance Coverage in the United States: 2014. Census.gov. Retrieved from https://www.census.gov/library/publications/2015/demo/P60-253.html

Solow, R.M. (1987). We'd better watch out. *New York Times Book Review*, 36.

Song, J. and Zahedi, F. (2007). Trust in health infomediaries. *Decision Support Systems*, 43(2), 390-407.

Sorlin, P. (2016). Sweden aims to become the world leader in eHealth. *eHealth Law & Policy*, 3(4). E-comlaw.com. Retrieved 11 October 2016, from http://www.e-comlaw.com/ehealth-law-and-policy/article_template.asp?Contents=Yes&from=ehlp&ID=280

Soti, P. and Pandey, S. (2007). Business process optimization for RHIOs. *Journal of Healthcare Information Management: JHIM*, 21(1), 40-47.

Spanakis, E.G., Kafetzopoulos, D., Yang, P., Marias, K., Deng, Z., Tsiknakis, M., Sakkalis, V. and Dong, F. (2014). MyHealthAvatar: Personalized and empowerment health services through Internet of Things technologies. *4th International Conference on Wireless Mobile Communication and Healthcare – Transforming Healthcare Through Innovations in Mobile and Wireless Technologies (MOBIHEALTH)*, 331-334. https://doi.org/10.1109/MOBIHEALTH.2014.7015978

Spanò, E., Di Pascoli, S. and Iannaccone, G. (2016). Low-power wearable ECG monitoring system for multiple-patient remote monitoring. *IEEE Sensors*, 16(13), 5452-5462. https://doi.org/10.1109/JSEN.2016.2564995

Spil, T. and Kijl, B. (2009). E-health business models: From pilot project to successful deployment. *IBIMA Business Review*, 1, 55-66.

Spil, T. and Klein, R. (2014, January). Personal Health Records Success: Why Google Health Failed and What Does that Mean for Microsoft Health Vault? pp. 2818-2827. *In: System Sciences (HICSS), 2014 47th Hawaii International Conference*. IEEE.

Squires, D. and Anderson, C. (2015). US health care from a global perspective: Spending, use of services, prices, and health in 13 countries. *Issue Brief (Commonwealth Fund)*, 15, 1-15.

Stansfield, S. (2008). An interview with Sally Stansfield – Who owns the information? Who has the power? *Bulletin of the World Health Organization*, 86(3).

Steele, E. (2006). EHR Implementation: Who Benefits, Who Pays? Health Management Technology. Retrieved 7 May 2014, from http://www.healthmgttech.com

Stephanie, L. (2011). Issues in the e-health ecosystem: Creating a digital commons. *Proceedings of the International Research Workshop on the Interactive Digital Media Marketplace, 2010*. Singapore.

Stephanie, L. (2017). S*ingapore's NEHR: Challenges on the path to connected health*. In Proceedings of 2017 IEEE International Conference on Industrial Engineering and Engineering Management (pp. 1128-1132). IEEE IEEM, Singapore, December 10-13.

Stephanie, L. and Sharma, R.S. (2010). Modeling digital flows in the e-health ecosystem: Strategic implications for players. *Proceedings of the Pacific Telecommunications Council Conference on Embracing the Cloud: Enabling Connectivity and Innovation, 2010*. Hawaii, USA.

Stephanie, L. and Sharma, R.S. (2011). Health on a cloud: Modeling digital flows in an e-health ecosystem. *2nd Symposium on Healthcare Advancements in Research and Practice (SHARP 2.0), 2011*. Minnesota, USA.

Stephanie, L. and Sharma, R.S. (2016). Health on a cloud: Modeling digital flows in an e-health ecosystem. *Journal of Advances in Management Sciences & Information System*, 2, 1-20.

Stephanie, L. and Sharma, R.S. (2019). Digital Health Eco-Systems: An epochal review of practice-oriented research. *International Journal of Information Management*, 102032. doi: 10.1016/j.ijinfomgt.2019.10.017

Stephanie, L., Sharma, R.S. and Ramasubbu, N. (2012). The digitization of Bollywood: Adapting to disruptive innovation. *Media Asia*, 39(1), 2012.

Stephanie, L., Srinivasan, T. and Lawale, A.D. (2011). Impact of Interactive Digital Media (IDM) on Healthcare. *In*: R.S. Sharma, M. Tan and F. Pereira (Eds.), Understanding the IDM Marketplace. Singapore: IGI Publishing.

Stephanie, L., Tan, M., Morales-Arroyo, M. and Sharma, R.S. (2011). The pervasiveness of mobile data services: Do usage and attitudinal divides exist in Asia and America? *International Journal of Electronic Business*, 9(1/2), 63-105. http://dx.doi.org/10.1504/ijeb.2011.040356

Steuer, J. (1992). Defining virtual reality: Dimensions determining telepresence. Communication, 42(4), 73-93. https://doi.org/10.1111/j.1460-2466.1992.tb00812.x

Stewart, D.W. and Kamins. M.A. (1993). *Secondary Research: Information Sources and Methods* (2nd ed.). California: Sage Publications, Inc.

Storf, H., Becker, M. and Riedl, M. (2009). Rule-based activity recognition framework: Challenges, technique and learning. *3d International ICST Conference on Pervasive Computing Technologies for Healthcare*. London, UK. https://doi.org/10.4108/ICST.PERVASIVEHEALTH2009.6108

Suddaby, R. (2006). From the Editors: What grounded theory is not. *Academy of Management Journal*, 49(4), 633-642. http://dx.doi.org/10.5465/amj.2006.22083020

Sujith, E. (2008). *Cloud Computing in Healthcare*. Frost & Sullivan Market Insight. Retrieved from https://www.frost.com/sublib/display-market-insight.do?id=135578323

Sun, F. (2018, May 7). A Survey of Consensus Algorithms in Crypto. *Medium*. https://medium.com/@sunflora98/a-survey-of-consensus-algorithms-in-crypto-e2e954dc9218

Sunyaev, A. and Schneider, S. (2013). Cloud services certification. *Communications of the ACM*, 56(2), 33-36.

Sunyaev, A., Chornyi, D., Mauro, C. and Krcmar, H. (2010, January). Evaluation framework for personal health records: Microsoft Health Vault vs. Google Health. pp. 1-10. *In*: *System Sciences (HICSS), 2010 43rd Hawaii International Conference on IEEE*.

Suppiah, V.L. (2016). Malaysia's healthcare industry to bring in RM13billion this year. *MIMS News*. Retrieved 8 November 2016, from http://today.mims.com/topic/malaysia-s-healthcare-industry-to-bring-in-rm13billion-this-year

Surhone, L.M., Timpledon, M.T. and Marseken, S.F. (2010). *Temporal Logic*. Betascript Publishing.

Suveen, A., Krumholz, H.M. and Schulz, W.L. (2017). Blockchain technology applications in health care. *Circulation: Cardiovascular Quality and Outcomes*, 10(9), 1-3. https://doi.org/10.1161/CIRCOUTCOMES.117.003800

Swan, M. (2015). *Blockchain: Blueprint for a New Economy* (1st ed.). O'Reilly Media.

Syeda-Mahmood, T.F., Walach, E., Beymer, D., Gilboa-Solomon, F., Moradi, M., Kisilev, P., Kakrania, D., Compas, C.B., Wang, H., Negahdar, M., Cao, Y., Baldwin, T., Guo, Y., Gur, Y., Rajan, D., Zlotnick, A., Rabinovici-Cohen, S., Ben-Ari, R., Amit, G., ... Hashoul, S.Y. (2016). Medical sieve: A cognitive assistant for radiologists and cardiologists. *Medical Imaging: Computer-Aided Diagnosis.* https://doi.org/10.1117/12.2217382

Tahir, D. (2014). Epic Systems feeling heat over interoperability. *Modern Healthcare.* Retrieved from http://www.modernhealthcare.com/article/20141001/NEWS/310019945

Talukder, A.K., Chaitanya, M., Arnold, D. and Sakurai, K. (2018). Proof of disease: A blockchain consensus protocol for accurate medical decisions and reducing the disease burden. *IEEE SmartWorld, Ubiquitous Intelligence & Computing, Advanced & Trusted Computing, Scalable Computing & Communications, Cloud & Big Data Computing, Internet of People and Smart City Innovation (SmartWorld/SCALCOM/UIC/ATC/CBDCom/IOP/SCI),* 257-262. https://doi.org/10.1109/SmartWorld.2018.00079

Tang, P.C., Ash, J.S., Bates, D.W., Overhage, J.M. and Sands, D.Z. (2006). Personal health records: Definitions, benefits, and strategies for overcoming barriers to adoption. *Journal of the American Medical Informatics Association,* 13(2), 121-126. Retrieved from http://search.proquest.com/docview/220828100?accountid=12665

The Office of the National Coordinator. (2010). Get the Facts about STATE HEALTH INFORMATION EXCHANGE PROGRAM. Retrieved 12 June 2012, from https://www.healthit.gov/sites/default/files/get-the-facts-about-state-hie-program-2.pdf

The Sequoia Project. (2017). eHealth Exchange Overview. Retrieved from http://sequoiaproject.org/ehealth-exchange/about/

Thomas, K. (2013). Q & A. *Healthcare Financial Management,* 67(11), 64-68. Retrieved from http://search.proquest.com/docview/1461712328?accountid=12665

Thomas, S.S., Nathan, V., Zong, C., Soundarapandian, K., Shi, X. and Jafari, R. (2016). BioWatch: A noninvasive wrist-based blood pressure monitor that incorporates training techniques for posture and subject variability. *Biomedical and Health Informatics,* 20(5), 1291-1300. https://doi.org/10.1109/JBHI.2015.2458779

Tongco, M. (2007). Purposive sampling as a tool for informant selection. *Ethnobotany Research and Applications,* 5, 147. http://dx.doi.org/10.17348/era.5.0.147-158

Top of Mind 2020: Patient Engagement, Data Analytics and Precision Medicine. (2020, September 16). Health IT Briefing. https://healthmanagement.org

Top of Mind for Top Health Systems 2020. (2020). Center for Connected Medicine (CCM) and KLAS Research. https://connectedmed.com/resources/top-of-mind-2020-data-analytics-precision-medicine-patient-engagement/

Torda, P., Han, E. and Scholle, S. (2010). Easing the adoption and use of electronic health records in small practices. *Health Affairs,* 29(4), 668-675. http://dx.doi.org/10.1377/hlthaff.2010.0188

Treskes, R.W., Van Der Velde, E.T., Atsma, D.E. and Schalij, M.J. (2016). Redesigning healthcare: The 2.4 billion euro question? *Netherlands Heart Journal,* 24(7), 441-446. doi:10.1007/s12471-016-0834-6

Tripathi, M., Delano, D., Lund, B. and Rudolph, L. (2009). Engaging patients for health information exchange. *Health Affairs,* 28(2), 435-443.

Truog, R. (2012). Patients and doctors—The evolution of a relationship. *New England Journal of Medicine,* 366(7), 581-585. http://dx.doi.org/10.1056/nejmp1110848

Tseng, J.C.C., Lin, B.-H., Lin, Y.-F., Tseng, V.S., Day, M.-L., Wang, S.-C., Lo, K.-R. and Yang, Y.-C. (2016). An interactive healthcare system with personalized diet and exercise guideline recommendation. 525-532. Scopus. https://doi.org/10.1109/TAAI.2015.7407106

Turocy, T.L. and von Stengel, B. (2001). Game Theory. CDAM Research Report LSE-CDAM-2001-09, October 2001.

Uddin, Md. A., Stranieri, A., Gondal, I. and Balasubramanian, V. (2018). Continuous patient monitoring with a patient centric agent: A block architecture. *IEEE Access*, 6, 32700-32726. https://doi.org/10.1109/ACCESS.2018.2846779

Urquhart, C. and Fernández, W. (2013). Using grounded theory method in information systems: The researcher as blank slate and other myths. *J. Inf. Technol*, 28(3), 224-236. http://dx.doi.org/10.1057/jit.2012.34

Urquhart, C., Lehmann, H. and Myers, M.D. (2010). Putting the 'theory' back into grounded theory: Gguidelines for grounded theory studies in information systems. *Information Systems Journal*, 20(4), 357-381. doi:10.1111/j.1365-2575.2009.00328.x

Van Limburg, M., Gemert-Pijnen, V., Nijland, N., Ossebaard, H.C., Hendrix, R.M.G. and Sevdel, E.R. (2011). Why business modeling is crucial in the development of e-health technologies. *Journal of Medical Internet Research*, 13(4), e124.

Varatharajan, R., Manogaran, G., Priyan, M.K. and Sundarasekar, R. (2018). Wearable sensor devices for early detection of Alzheimer disease using dynamic time warping algorithm. *Cluster Computing*, 21(1), 681-690. https://doi.org/10.1007/s10586-017-0977-2

Veltink, P.H., Bussmann, Hans B.J., de Vries, W., Martens, Wim L.J. and Van Lummel, R.C. (1996). Detection of static and dynamic activities using uniaxial accelerometers. *IEEE Transactions on Rehabilitation Engineering*, 4(4), 375-385. https://doi.org/10.1109/86.547939

Velu, J.F., Groot Jebbink, E., de Vries, J.-P.P.M., Slump, C.H. and Geelkerken, R.H. (2017). Validation of the Simbionix PROcedure Rehearsal Studio sizing module: A comparison of software for endovascular aneurysm repair sizing and planning. *Vascular*, 25(1), 80-85. https://doi.org/10.1177/1708538116651009

Vest, J.R., Campion Jr, T.R. and Kaushal, R. (2013). Challenges, alternatives, and paths to sustainability for health information exchange efforts. *Journal of Medical Systems*, 37(6), 1-8.

Vigna, P. and Casey, M.J. (2016). *The Age of Cryptocurrency: How Bitcoin and the Blockchain Are Challenging the Global Economic Order* (Reprint edition). Picador.

Vishwanath, A. and Scamurra, S.D. (2007). Barriers to the adoption of electronic health records: Using concept mapping to develop a comprehensive empirical model. *Health Informatics Journal*, 13(2), 119-134.

Vishwanath, A. and Scamurra, S.D. (2007). Barriers to the adoption of electronic health records: Using concept mapping to develop a comprehensive empirical model. *Health Informatics Journal*, 13(2), 119-134.

Vitalari, N.P. (2015). A Prospective Analysis of the Future of the US Healthcare Industry.

Von Neumann, J. and Morgenstern, O. (1953). *Theory of Games and Economic Behavior* (3rd Ed.). Princeton University Press.

Wagner, M. (2009). Keas Puts Healthcare Advice, Apps Online. *Information Week*. Retrieved from http://www.informationweek.com/news/healthcare/interoperability/showArticle.jhtml?articleID=220600350

Wahab, A. and Memood, W. (2018). *Survey of Consensus Protocols*. 12.

References

Wainer, H. and Braun, H.I. (1988). *Test Validity*. Hillsdale, NJ: Lawrence Erlbaum Associates.

Walker, J., Pan, E., Johnston, D., Adler-Milstein, J., Bates, D. and Middleton, B. (2005). The Value of Health Care Information Exchange and Interoperability. *Health Affairs*, 24, 10-18.

Wan, J., A.A.H. Al-awlaqi, M., Li, M., O'Grady, M., Gu, X., Wang, J. and Cao, N. (2018). Wearable IoT enabled real-time health monitoring system. *Wireless Communications and Networking (EURASIP), 2018*(1), 298. https://doi.org/10.1186/s13638-018-1308-x

Wan, Jie, Li, M., O'Grady, M., Gu, X., A.A.H. Alawlaqi, M. and O'Hare, G. (2018). Time-bounded activity recognition for ambient assisted living. *IEEE Transactions on Emerging Topics in Computing*, 1-1. https://doi.org/10.1109/TETC.2018.2870047

Wang, H. and Song, Y. (2018). Secure cloud-based EHR system using attribute-based cryptosystem and blockchain. *Medical Systems*, 42(8), 152. https://doi.org/10.1007/s10916-018-0994-6

Wang, S., Wang, J., Wang, X., Qiu, T., Yuan, Y., Ouyang, L., Guo, Y. and Wang, F.-Y. (2018). Blockchain-powered parallel healthcare systems based on the ACP approach. *IEEE Transactions on Computational Social Systems*, 5(4), 942-950. https://doi.org/10.1109/TCSS.2018.2865526

Wang, Y. and Hajli, N. (2017). Exploring the path to big data analytics success in healthcare. *Journal of Business Research*, 70, 287-299. http://dx.doi.org/10.1016/j.jbusres.2016.08.002

Webb, G. (2012). Making the cloud work for healthcare (cover story). *Health Management Technology*, 33(2), 8-9.

Webster, J. and Watson, R.T. (2002). Analyzing the past to prepare for the future: Writing a literature review (guest editorial). *MIS Quarterly*, 26(2), xiii-xxiii.

Weiss, P.L., Kizony, R., Feintuch, U. and Katz, N. (2006, February). Virtual reality in neurorehabilitation. *Textbook of Neural Repair and Rehabilitation*. https://doi.org/10.1017/CBO9780511545078.015

Wen, H.J. and Tan, J. (2003, January). The evolving face of telemedicine and e-health: Opening doors and closing gaps in e-health services opportunities and challenges. *In: System Sciences, 2003*. Proceedings of the 36th Annual Hawaii International Conference on System Sciences (HICSS'03). IEEE.

Whitten, P., Steinfield, C. and Hellmich, S. (2001). E-health: Market potential and business strategies. *Journal of Computer-Mediated Communication*, 6(4).

Wickramasinghe, N.S., Fadlalla, A.M.A., Geisler, E. and Schaffer, J.L. (2005). A framework for assessing e-health preparedness. *International Journal of Electronic Healthcare*, 1(3), 316-334.

Wiedemann, L.A. (2012). A look at unintended consequences of EHRs. *Health Management Technology*, February 2012, 24-25.

Williams, C. (2007). Research Methods. *Journal of Business & Economic Research*, 5(3), 65-72.

Williams, M.S., Buchanan, A.H., Davis, F.D., Faucett, W.A., Hallquist, M.L.G., Leader, J.B., Martin, C.L., McCormick, C.Z., Meyer, M.N., Murray, M.F., Rahm, A.K., Schwartz, M.L.B., Sturm, A.C., Wagner, J.K., Williams, J.L., Willard, H.F. and Ledbetter, D.H. (2018). Patient-centered precision health in a learning health care system: Geisinger's genomic medicine experience. *Health Affairs*, 37(5), 757-764. https://doi.org/10.1377/hlthaff.2017.1557

World Health Statistics 2016: Monitoring health for the SDGs. (2016). World Health

Organization. Retrieved 8 November 2016, from http://www.who.int/gho/publications/world_health_statistics/2016/dashboard/en/

Wu, H.-K., Lee, S.W.-Y., Chang, H.-Y. and Liang, J.-C. (2013). Current status, opportunities and challenges of augmented reality in education. *Computers & Education*, 62, 41-49. https://doi.org/10.1016/j.compedu.2012.10.024

Wu, W., Zhang, H., Pirbhulal, S., Mukhopadhyay, S.C. and Zhang, Y.-T. (2015). Assessment of biofeedback training for emotion management through wearable textile physiological monitoring system. *IEEE Sensors*, 15(12), 7087-7095. https://doi.org/10.1109/JSEN.2015.2470638

www.orcam.com. (n.d.). OrCam. Retrieved October 29, 2019, from https://www.orcam.com/en/

Xia, Q., Sifah, E.B., Asamoah, K.O., Gao, J., Du, X. and Guizani, M. (2017). MeDShare: Trust-less medical data sharing among cloud service providers via blockchain. *IEEE Access*, 5, 14757-14767. https://doi.org/10.1109/ACCESS.2017.2730843

Xia, Q., Sifah, E.B., Smahi, A., Amofa, S. and Zhang, X. (2017). BBDS: Blockchain-based data sharing for electronic medical records in cloud environments. *Information*, 8(2), 44. https://doi.org/10.3390/info8020044

Xu, J., Song, L., Xu, J.Y., Pottie, G.J. and van der Schaar, M. (2016). Personalized active learning for activity classification using wireless wearable sensors. *IEEE Journal of Selected Topics in Signal Processing*, 10(5), 865-876. https://doi.org/10.1109/JSTSP.2016.2553648

YahooNews, (2015). Four die as Singapore hospital suffers wave of hepatitis C infections. News.yahoo.com. Retrieved from http://news.yahoo.com/four-die-singapore-hospital-suffers-wave-hepatitis-c-175824102.html YahooNews, (2016) "can only meet one-fourth of the projected demand for hospital beds by 2030"

Yang, H. and Yang, B. (2018). A blockchain-based approach to the secure sharing of healthcare data. *Norwegian Information Security Conference*, 12.

Yaraghi, N. (2015). A Sustainable Business Model for Health Information Exchange Platforms: The Solution to Interoperability in Health Care IT. The Brookings Institution. Retrieved from http://www.brookings.edu/research/papers/2015/01/30-sustainable-business-model-health-information-exchange-yaraghi

Yarwood-Ross, L. and Jack, K. (2015). Using extant literature in a grounded theory study: A personal account. *Nurse Researcher*, 22(4), 18-24. http://dx.doi.org/10.7748/nr.22.4.18.e1316

Yellowlees, P.M., Marks, S.L., Hogarth, M. and Turner, S. (2008). Standards-based, open-source electronic health record systems: A desirable future for the U.S. health industry. *Telemedicine and e-Health*, 14(3), 284-288. doi: 10.1089/tmj.2007.0052

Yin, R.K. (2003). *Case Study Research: Design and Methods* (3rd ed.). Thousand Oaks: Sage Publications.

Yip, W. and Hsiao, W. (2009). China's health care reform: A tentative assessment. *China Economic Review*, 20(4), 613-619. http://dx.doi.org/10.1016/j.chieco.2009.08.003

Yoon, J.W., Chen, R.E., Han, P.K., Si, P., Freeman, W.D. and Pirris, S.M. (2017). Technical feasibility and safety of an intraoperative head-up display device during spine instrumentation. *Medical Robotics Computer Assisted Surgery (MRCAS)*, 13(3). https://doi.org/10.1002/rcs.1770

Young, B., Clark, C., Kansky, J. and Pupo, E. (2014). Definition: Continuum of Care | HIMSS. Himss.org. Retrieved from http://www.himss.org/ResourceLibrary/genResourceDetailPDF.aspx?ItemNumber=30272

References

Yudkowsky, R., Luciano, C., Banerjee, P., Schwartz, A., Alaraj, A., Lemole, G.M., Charbel, F., Smith, K., Rizzi, S., Byrne, R., Bendok, B. and Frim, D. (2013). Practice on an augmented reality/haptic simulator and library of virtual brains improves residents' ability to perform a ventriculostomy. *Society for Simulation in Healthcare*, 8(1), 25-31. https://doi.org/10.1097/SIH.0b013e3182662c69

Yue, X., Wang, H., Jin, D., Li, M. and Jiang, W. (2016). Healthcare data gateways: Found healthcare intelligence on blockchain with novel privacy risk control. *Medical Systems*, 40(10), 218. https://doi.org/10.1007/s10916-016-0574-6

Zahedi, F. and Song, J. (2008). Dynamics of trust revision: Using health infomediaries. *Journal of Management Information Systems*, 24(4), 225-248.

Zauderer, M.G., Gucalp, A., Epstein, A.S., Seidman, A.D., Caroline, A., Granovsky, S., Fu, J., Keesing, J., Lewis, S., Co, H., Petri, J., Megerian, M., Eggebraaten, T., Bach, P. and Kris, M.G. (2014). Piloting IBM Watson Oncology within Memorial Sloan Kettering's regional network. *Clinical Oncology*, 32(15_suppl), e17653–e17653. https://doi.org/10.1200/jco.2014.32.15_suppl.e17653

Zhang, A. and Lin, X. (2018). Towards secure and privacy-preserving data sharing in e-health systems via consortium blockchain. *Medical Systems*, 42(8), 140. https://doi.org/10.1007/s10916-018-0995-5

Zhang, C., Wu, J., Long, C. and Cheng, M. (2017). Review of existing peer-to-peer energy trading projects. *Energy Procedia*, 105, 2563-2568. https://doi.org/10.1016/j.egypro.2017.03.737

Zhang, F., Cecchetti, E., Croman, K., Juels, A. and Shi, E. (2016). *Town Crier: An Authenticated Data Feed for Smart Contracts* (No. 168). http://eprint.iacr.org/2016/168

Zhang, J., Xue, N. and Huang, X. (2016). A secure system for pervasive social network-based healthcare. *IEEE Access*, 4, 9239-9250. https://doi.org/10.1109/ACCESS.2016.2645904

Zhang, P., White, J., Schmidt, D.C., Lenz, G. and Rosenbloom, S.T. (2018). FHIRChain: Applying blockchain to securely and scalably share clinical data. *Computational and Structural Biotechnology*, 16, 267-278. https://doi.org/10.1016/j.csbj.2018.07.004

Zhang, Y., Qiu, M., Tsai, C.-W., Hassan, M.M. and Alamri, A. (2017). Health-CPS: Healthcare cyber-physical system assisted by Cloud and Big Data. *IEEE Systems*, 11(1), 88-95. https://doi.org/10.1109/JSYST.2015.2460747

Zhao, H., Zhang, Y., Peng, Y. and Xu, R. (2017). Lightweight backup and efficient recovery scheme for health blockchain keys. *13th International Symposium on Autonomous Decentralized System (ISADS)*, 229-234. https://doi.org/10.1109/ISADS.2017.22

Zheng, Z., Xie, S., Dai, H., Chen, X. and Wang, H. (2017). An overview of blockchain technology: Architecture, consensus, and future trends. *IEEE International Congress on Big Data (BigData Congress)*, 557-564. https://doi.org/10.1109/BigDataCongress.2017.85

Zheng, Z., Xie, S., Dai, H.N., Chen, X. and Wang, H. (2018). Blockchain challenges and opportunities: A survey. *Web and Grid Services*, 14(4), 352. https://doi.org/10.1504/IJWGS.2018.10016848

Zhu, Licai, Wang, R., Wang, Z. and Yang, H. (2017). TagCare: Using RFIDs to monitor the status of the elderly living alone. *IEEE Access*, 5, 11364-11373. https://doi.org/10.1109/ACCESS.2017.2716359

Zhu, Liehuang, Wu, Y., Gai, K. and Choo, K.-K.R. (2019). Controllable and trustworthy blockchain-based cloud data management. *Future Generation Computer Systems*, 91, 527-535. https://doi.org/10.1016/j.future.2018.09.019

Zieth, C.R., Chia, L.R., Roberts, M.S., Fischer, G.S., Clark, S., Weimer, M. and Hess, R. (2014). The evolution, use, and effects of integrated personal health records: A narrative review. *Electronic Journal of Health Informatics*, 8(2), e17.

Zinberg, J. (2016). Are provider-led health care networks too big to fail? *Health Affairs*. Retrieved from http://healthaffairs.org/blog/2016/01/12/are-provider-led-health-care-networks-too-big-to-fail/

Zott, C. and Amit, R. (2010). Business model design: An activity system perspective. *Long Range Planning*, 43(2-3), 216-226. http://dx.doi.org/10.1016/j.lrp.2009.07.004

Zyskind, G., Nathan, O. and Pentland, A. "Sandy". (2015). Decentralizing privacy: Using blockchain to protect personal data. *IEEE Security and Privacy Workshops*, 180-184. https://doi.org/10.1109/SPW.2015.27

Index

3M scheme, 104

A

ACA, 124
Accountable Care Organizations (ACO), 129
Adoption by consumers, 89
ADVISOR, 87, 89, 90, 102
Ageing population, 106
Agency for Integrated Care, 106
Aggregation of patient data, 127
Aggregation, 87
Alternative healthcare practitioners, 84
American hospitals, 120
American Recovery and Reinvestment Act (ARRA), 123
Analytics and informatics, 148
Anonymization, 112
Artificial intelligence, 14, 36, 39, 47
Assessment tools, 23, 37
Asset class, 148

B

Bankruptcies, 122
Bankruptcy of General Motors, 123
Big data analytics, 112, 152
Big Data, 18, 36, 39-42, 45-47, 50, 127, 148
Bitcoin, 27-28, 33
Blockchain, 14, 18-20, 25-35
Business Information (BI) System, 112
Business model, 58, 62

C

Carrot and stick approach, 124, 156
Case study, 8
Certification Commission for Healthcare Information Technology (CCHIT), 82
Certification, 81
Children's Health Insurance Program (CHIP), 122
Citizen-centric value, 160
Citizen-centric, 157
Clear mandate, 156
CLEO, 110
Clinic Electronic Medical Record and Operation System, 110
Clinical decision making, 148
Clinical Decision Support (CDS), 125
Closed records, 134
Cloud computing, 61
Cloud-based infomediaries, 80
Coding reliability, 69
Coding scheme, 76
Commons, 56
Communication network providers, 153
Computerized Physician Order Entry (CPOE), 125
Concept, 26
Conceptual model, 64
Confidentiality, 81
Connectedness, 155
Consensus algorithm, 27, 28
Consumer awareness and choice, 151
Consumer-centrism, 147
Consumerism, 4, 93

238 *Index*

Continuum of Care Document
  Architecture (CCDA), 115
Continuum of care, 110, 155
Cooperation dilemma, 56, 116
Cost-efficiency, 85, 87
Critical success factor framework, 157
Critical success factors, 138
Cross-case analysis, 76, 139
Cross-case comparison, 77
Cut-backs, 138

D

Data connectivity, 12
Data management, 28-29, 33-34, 41
Data storage and security, 12, 18, 25
DataCloud award, 150
Descriptive case studies, 63
Digital exchanges, 88
Digital Healthcare, 12, 28-34
Digital twin, 36, 46-47
Disruptive innovation, 89, 152
Distribution, 87

E

Economic sustainability, 138
Economies of scope, 111
Effective, 100
Efficiency, 54, 99, 101
Efficient, 95, 100
e-health adoption game, 145
e-health business models, 6
eHealth Exchange, 127, 148, 151, 156
e-health network, 4
e-health preparedness, 104
e-health, 2, 52, 53, 61
EHR adoption, 5
EHR Incentive Program, 124, 137
EHR-integrated PHR, 93
Electronic Health Intelligence System
  (eHINTS), 112
Electronic Health Records (EHR), 3, 62
Electronic Medical Record Exchange
  (EMRX), 105
Electronic Medical Records (EMR), 61
Emerging information technologies, 12
Empowerment, 135
Encryption, 136

Entrepreneurial behavior, 111
Episodic care, 126
e-revolution, 1
Evidence-based medicine, 94
e-visits, 135
expert interviews, 78
Exploratory case studies, 63
Exponential, 150

F

Fair, 95
Fairness, 54, 61, 99, 101
Fee-for-service, 111
Fragmented healthcare system, 155
Fragmented state, 123
Free-riding, 62, 115
Futuristic e-health ecosystem, 157

G

Game theory, 53, 61
Google Health, 80
Government intervention, 156

H

H-Cloud, 108
Health analytics, 59
Health Information Exchange (HIE), 4,
  124
Health information technology (HIT), 2
Health Information Technology for
  Economic and Clinical Health
  (HITECH), 9, 52, 75, 120
Health Information Technology for
  Economic and Clinical Health Act,
  123, 124
Health Insurance Exchange Marketplace,
  124
Health Insurance Portability and
  Accountability Act (HIPAA), 82, 123
Health Level Seven (HL7), 82
Healthcare consumerism, 151
Healthcare Information Management
  Systems & Society, 110
HealthHub, 114, 151
Healthiest country, 104
HER, 53
HIE (Health Information Exchange), 79

Index **239**

HIT vendors, 136
HSOR (Health Services & Outcomes
   Research), 113

I

Image sensors, 22
Implementations, 47-51
Improved services, 85, 87
Independent PHR, 152
Inertial sensors, 21
Infomediaries, 79, 82, 86, 94, 100
Infomediary, 59, 61
Informants, 76
Informatics, 12
Information asymmetry, 56, 62, 134
Information blocking, 57, 62
Information flows, 88
Information richness, 72, 77, 140
Integrated care, 126
Integrated delivery systems, 131
Integrated Health Information Systems,
   108
Integrated health system, 110
Integrity, 81
Inter-cluster competition, 105
Interface, 89
Intermediary, 79
Internet of Things, 12
Interoperability, 20, 34, 61, 95, 137
Intra-organizational benefits, 147
Investment burden, 85

K

Key players, 8

L

Lifelong story, 142
Location sensors, 22
LOINC (Logical Observation Identifiers
   Names and Codes), 82
Longitudinal health record, 155

M

Misaligned incentives, 128
Machine learning, 31-32, 36-37
Managed Care Organizations (MCO), 129

Maturity curve, 126
Maximum variation, 72, 77, 140
Medicaid, 122
Medical errors, 121
Medicare, 122
Microsoft Health Vault, 80
Mobile health apps, 153
Modelling, 101
MOH Holdings, 107

N

Narrow network, 116, 134, 147
National architecture, 151
National Electronic Health Records
   (NEHR), 9, 52, 75, 104, 107
National Healthcare Group, 109
National level infomediary, 127
National scale, 140
NCPDP (National Council for
   Prescription Drug Programs), 82
Network effects, 111
Network participation fees, 127
Neutral private entity, 148
Non-cooperative games, 144

O

ObamaCare, 124
Office of the National Coordinator for
   Health Information Technology
   (ONCHIT), 123
One Patient One Record, 106, 147
Optimal ecosystem, 143
Organizing model, 89
Overview, 12, 14, 28, 51

P

Paradigm shift, 126, 147
Pareto-efficiency, 61, 146
Pareto-efficient, 54, 55, 101, 103
Participation dilemmas, 55, 115
Patient autonomy (empowerment), 81
Patient Health Records (PHR), 131
Patient Protection and Affordable Care
   Act, 124
Patient-centric e-health, 3, 6, 10, 77
Patient-centric, 58, 61

Patients (Consumers), Providers, Payers, Vendors, Infomediaries and Regulators, 8
Patients (Healthcare Consumers), 82, 86
Patients, 79
Patients/Consumers, 84
Pay for Performance (P4P), 130
Payers, 79, 82, 86
Pay-for-performance, 111
Payoff graph, 144
Payoffs, 10
Personal Data Protection Act (PDPA), 113
Personal Health Records (PHR), 53, 61
Personalized healthcare, 13-15, 22-23, 35-45
Pharmaceutical companies, 153
Physiological sensors, 22
Pluralistic, 132
Population health management, 148
Population-centric healthcare, 119
Population-centric, 110
Precision healthcare, 35, 47
Preventive healthcare, 108, 159
Prisoner's dilemma, 55, 62
Privacy, 81
Private cloud, 109
Productivity paradox, 56, 115, 133
Provider-centric, 58, 61
Provider-centrism, 57
Provider-led healthcare networks, 131
Providers, 79, 82, 86
Public good, 99
Public-private healthcare, 140
Public-private sector, 121
Public-private, 148
Purposive sampling, 60

Q

Qualitative research, 70, 71

R

Readmission rates, 111
Recognized interoperability standards, 155
Regional Health Information Organizations (RHIO), 127

Regional Health Systems, 106
Regulators, 80, 82, 86
Remote monitoring, 14, 33
Research questions, 6
Research roadmap, 64
Revenue/Cost sharing, 89

S

Secondary research, 70
Security and privacy, 30, 33-34
Security, 81
Semi-structured interviews, 71
Service platform, 89
Silo, 134
Silver tsunami, 106, 149, 159
SingHealth, 109
Smart contracts, 14, 18, 25-31
Smart Homes, 108
Smart Mat, 108
SNOMED (The Systematized Nomenclature of Medicine), 82
Social capital, 159
Sophisticated healthcare system, 121
Stable, 95
Standardization, 112
Standards, 81
Structured interviews, 71
Sustainable Development Goals (SDGs), 160
Sustainable patient-centric e-health business model, 10
Sustainable, 54, 95, 153
Syndication, 87

T

Taxonomy, 36
Thematic literature review, 65
Theoretical framework, 140
Third-party vendors, 86
To Err is Human, 121
Total digital health system (TDHS), 5, 7
Tragedy of the commons, 3, 62
Tragedy of the digital commons, 56, 133
Transaction flows, 88
Two-sided market, 117
Two-tier system, 104

Index 241

U

Ubiquitous access, 142
Ubiquitous, 135
Unaffordability, 122
Unit of analysis, 72
Universal health coverage, 149
Unstructured interviews, 71

V

Value capture, 62, 67
Value creation, 62, 67, 94, 95

Value proposition, 89
Values captured, 99
Values created, 99
Vendors, 79, 82
Venn diagram, 141

W

Wearable healthcare devices, 153
Wearable sensors, 12-13, 15-16, 21

Milton Keynes UK
Ingram Content Group UK Ltd.
UKHW050641210624
444299UK00002B/5